'Plesner and Husted provide a refreshing perspective o
today's organizations and ways of working. They remir
in what already exists and that technology and or
intertwined. It is a timely and useful textbook for stu
Fayard, *Associate Professor Innovation, Design and Orga
of Engineering, USA*

'A brilliant, up-to-date and inspiring book: a must-read for every scholar interested in the relationship between technologies and work practices, digitalizing and organizing processes. Sometimes a textbook is needed, and sometimes a textbook does more than one would expect. This is the case.' – **Attila Bruni**, *Associate Professor, Department of Sociology and Social Research, Trento University, Italy*

'*Digital Organizing* is the textbook I had been desperately looking for – even before I knew it existed. Today's students in management are keen to learn all that is new about the digital economy, yet management academics have a responsibility to take them beyond the contemporary hype, pointing to the enduring relevance of existing organizational theories. Plesner and Husted achieve both goals, masterfully, by revisiting organizational theory and locating it in a digital context.' – **Dr Daniel Beunza**, *Associate Professor of Management, Cass Business School, UK*

'Starting with algorithmic thinking in the nineteenth century and moving through a multitude of cases of organizing the digital, this text offers a thorough grounding in contemporary matters of concern for organization studies. By refusing to get drawn in by alluring technological hyperbole, the authors consider the digital through well-established organizational themes such as knowledge, power and structure. The end result is a thought-provoking text, challenging us to rethink how we do organizational studies of the digital.' – **Professor Daniel Neyland**, *Head of Sociology, Goldsmiths, University of London*

'This book places organisations at the heart of digitalisation. It revisits classic themes in organizations studies in order to elicit the continuities and discontinuities embedded in digital organising. The book takes a much sought-after critical approach to how the digital intertwines with the organisational and provides readers with a vocabulary to understand and illustrate this co-constitutive relationship. This is a book that must be read by all those who study, teach and research on technology and organization studies.' – **Dr Dimitra Petrakaki**, *Reader in Information Systems (Management), University of Sussex, UK*

'The typewriter was revolutionary in its day, replacing direct control with mediating bureaucracy embedded in written rules. Today, the new disruption comes from the digitalization of practices that were formerly more bureaucratically embedded. For many proponents of the digital, it changes everything. This book is more balanced, seeing the digital revolution not only in its discontinuities with past practices but also its continuities. As an antidote to hype, it will be an invaluable resource.' – **Stewart Clegg**, *Distinguished Professor, University of Technology Sydney Business School, Australia*

'I highly recommend Plesner and Husted's book. It provokes students, teachers, researchers and practitioners to think about digitalization organizationally. The book is a comprehensive source of information on current debates in organisation studies on the role and impact of digital technologies. It will be a staple in my teaching and research.' – **Hannah Trittin-Ulbrich**, *Assistant Professor for Business Ethics, Leuphana University of Lüneburg, Germany*

'In contrast to speculative musings about the future of work, this excellent book provides a profound and systematic overview of how work is actually changing with digital technologies. Uniquely, the book's approach is to think 'organizationally' about digitalization rather than focusing on technology per se. It is a must read for anyone involved in organization studies.' – **Judy Wajcman**, *Anthony Giddens Professor of Sociology, London School of Economics, UK*

'With a fresh digital angle, this book provides an exhaustive overview of some of the most relevant topics in organization studies. This re-interpretation and analysis of organizational challenges and processes is accomplished with concise frameworks that make the book an accessible entry point for both scholars and students, who are interested in all sorts of matters related to contemporary digitized organizations.' – **Michael Etter**, *Senior Lecturer in Entrepreneurship and Digitization, King's College London, UK*

'With *Digital Organizing*, organization studies – historically a technical science – rediscovers technology as a fundamental aspect of its analysis. It presents a systematic account of ways to theoretically conceptualize and empirically analyse the digital. It shows that the digital is not simply transforming organizations, but is itself organizationally conditioned. The analysis is historically embedded and drawn from empirical case studies, e.g. digitalizing libraries, journalism, health care, social media, public relations and self-employment. It is a much-needed contribution that presents us with a stark contrast to the mostly sensationalist and utopian literature that dominates discussions of processes of digitalization.' – **Dr Robert Seyfert**, *University of Duisburg-Essen, Germany*

'A truly engaging, well-crafted and essential contribution! Highly recommended reading for everyone interested in the role of organization in the digital age.' – **Dr Stephan Schaefer**, *Researcher and Lecturer in Organization Studies, Lund University, Sweden*

'In today's constantly-changing digitally-enabled environment, managers need a structured guide to innovation that does justice to the breadth of the topic. *Managing Digital Innovation* delivers this really well, covering not just innovation itself, but also strategy, organisational structures, project management and the many relationships between knowledge and innovative activities. A case-based approach featuring organisations from many different countries combines theoretical rigour with the practical implications that current and future managers need to understand.' – **Professor John S Edwards**, *Aston Business School, UK*

'*Digital Organizing* is a well-written, balanced and insightful textbook. It accomplishes what has proved difficult for others: addressing the digital landscape in which contemporary organizations operate without arguing that everything has changed and that 'old' theories of organizing have become obsolete yet at the same time pointing to concrete transformations that require our attention and subsequently new theoretical approaches.' – **Ib T. Gulbrandsen**, *Associate Professor, Roskilde University, Denmark*

'The relationship between organization and (digital) technologies has been strangely neglected in Organization Studies in recent times, despite significant changes in society. Consequently, this book could not be timelier. Cleverly interweaving classical OS concepts with emerging digitalisation practices – for example, collaboration and co-creation, knowledge and datafication – Plesner and Husted's book is an invaluable resource for a wide range of scholars who are seeking to understand digitalisation, work and organization in the 21st century.' – **Gillian Symon**, *Professor of Organization Studies and Co-Director Digital Organization and Society Research Centre, Royal Holloway, University of London*

'This book spells out organizational thinking about digital media technologies in a sophisticated yet accessible way. It is urgently needed: it reminds us of key organizational issues and counteracts some of the hype surrounding digital media technologies. It provides a very helpful and insightful bridge between debates in organization studies and fields such as science and technology studies, and shows what organizational thinking can contribute to tackling some of the challenges of the digital transformation.' – **Professor Armin Beverungen**, *Leuphana University Lüneburg, Germany*

'*Digital Organizing* is a valuable and timely contribution to the study of digitalization and organizational life. By revisiting classical themes in organization studies, Plesner and Husted establish a much-needed vocabulary for thinking organizationally about digitalization. The clear approach to different concepts and perspectives makes *Digital Organizing* the perfect companion for teaching about the digital in management, organization studies and related disciplines.' – **Cristina Alaimo**, *Assistant Professor, Surrey Business School, University of Surrey, UK*

'This book provides an informed and instructive account of how digitization is organizationally mediated, and so is made intelligible in relation to diverse organizational phenomena. Its accessibility enables students of organization studies to appreciate the significance, but also the limits and pathologies, of digitization in a range of organizational contexts. It also reflects the wider implications of digitization, such as its influence upon the reconstitution of the division of labour.' – **Hugh Willmott**, *Professor of Management, Cass Business School, UK*

'Digital technologies are an inescapable dimension of our lives and have dramatically changed how work is done. This book is both passionate and astute about the latest digital developments and helps us understand how these affect organizational life. Plesner and Husted provide not only a thorough approach to organizational studies but also an exciting analysis of digitalization. The book offers innovative resources for students of organization and technology while suffusing with interesting and helpful insights.' – **Oana Albu**, *Associate Professor in the Department of Management and Marketing, University of Southern Denmark, Denmark*

DIGITAL ORGANIZING

Revisiting Themes in Organization Studies

URSULA PLESNER AND
EMIL HUSTED

macmillan international HIGHER EDUCATION

RED GLOBE PRESS

© Ursula Plesner and Emil Husted, under exclusive licence to Springer Nature Limited 2020

All rights reserved. No reproduction, copy or transmission of this publication may be made without written permission.

No portion of this publication may be reproduced, copied or transmitted save with written permission or in accordance with the provisions of the Copyright, Designs and Patents Act 1988, or under the terms of any licence permitting limited copying issued by the Copyright Licensing Agency, Saffron House, 6-10 Kirby Street, London EC1N 8TS.

Any person who does any unauthorized act in relation to this publication may be liable to criminal prosecution and civil claims for damages.

The authors have asserted their rights to be identified as the authors of this work in accordance with the Copyright, Designs and Patents Act 1988.

First published 2020 by
RED GLOBE PRESS

Image editing and graphic illustrations: Mette Plesner

Red Globe Press in the UK is an imprint of Springer Nature Limited, registered in England, company number 785998, of 4 Crinan Street, London N1 9XW.

Red Globe Press® is a registered trademark in the United States, the United Kingdom, Europe and other countries.

ISBN 978-1-137-60491-0 paperback

This book is printed on paper suitable for recycling and made from fully managed and sustained forest sources. Logging, pulping and manufacturing processes are expected to conform to the environmental regulations of the country of origin.

A catalogue record for this book is available from the British Library.

A catalog record for this book is available from the Library of Congress.

CONTENTS

List of images, figures and tables — vi
Preface — viii

Part I: Digital organizing — 1

1. The digital and the organizational — 3
2. Technology as a theme in organization studies — 30
3. Perspectives on technology and organization — 53

Part II: Revisiting themes in organization studies — 79

4. Structure and infrastructure — 81
5. Production and produsage — 106
6. Collaboration and co-creation — 131
7. Knowledge and datafication — 153
8. Communication and interactivity — 174
9. Legitimacy and transparency — 196
10. Power and empowerment — 217
11. Implications of digital organizing — 240

Index — 258

LIST OF IMAGES, FIGURES AND TABLES

Images

1.1	Computer scientist Grace Hopper, standing at a computer in 1952	5
1.2	Old computer punch card	6
1.3	Connectivity	12
1.4	A robot priest wearing a Buddhist robe	18
1.5	A New York Stock Exchange screen	20
2.1	The interior of a gold stamp mill in 1888	33
3.1	Alexander Graham Bell with one of the first telephones	63
4.1	Paper files	86
4.2	Work in a shared office space around new digital platforms	91
4.3	An Occupy Wall Street demonstration	101
5.1	Pyramide capitaliste – a propaganda poster from 1911	111
5.2	Work with a 3D printer	121
5.3	Workers produce Chinese-made handsets in Guangdong province	126
6.1	Collaboration in a knowledge-intensive environment	138
6.2	The façade of the NYSE in 2012 on the occasion of the public offering of YELP	143
7.1	Handheld devices have become more common in home care	156
7.2	Baseball has become a datafied sport	164
8.1	Mobile devices allow us to be 'always on'	187
9.1	Our digital devices leave traces even when we move through the city	207
9.2	Crowdsourcing platforms challenge organizations by exposing them to uncensored criticism. Organizations may try to speak back to them in various ways...	212
10.1	Freelance platforms are new sites for trading one's labor	219
10.2	A Cuban prison built as a panopticon	227
11.1	Many people take advantage of all the digital technologies that measure and track performance	244
11.2	The practice of hacking makes cyber security an increasingly important issue	248

Figures

1.1	Elements in the development of 'the digital'	15
2.1	The industrial plant as a social system	35
2.2	The organization as an open system, as conceived by contingency theorists	41

LIST OF IMAGES, FIGURES AND TABLES

2.3	Organizational technologies categorized according to task analyzability and task variability	42
2.4	Notable contributions on technology as a theme in organization studies	49
3.1	Aspects of the affordance concept	72
4.1	Example of traditional organizational diagram	85
4.2	Example of network diagram	90
4.3	Example of flow diagram	92
4.4	Illustration of the virtual organization based on online collaboration	95
5.1	The traditional (industrial) production–consumption process	118
5.2	The prosumption process	119
5.3	The produsage process	119
6.1	Phases in collaborative arrangements	137
7.1	The relationship between knowledge and data	163

Tables

1.1	The concepts of digitization and digitalization	7
2.1	Equivoque technologies	46
3.1	Three perspectives on technology and organization	68
6.1	Dimensions of digital platforms for collaboration	147
8.1	Perspectives on communication in organizations	191
9.1	Models of corporate legitimacy	199
9.2	Methods for manufacturing transparency	210
10.1	Functions of digital technologies in the workplace	229
10.2	The four faces of power in organizations	234
11.1	Dilemmas of digital organizing	254

PREFACE

Walk into any conventional bookstore or browse through Amazon's endless catalogue of catchy titles, and you will find countless textbooks promising to reveal how digital technology has overwhelmed contemporary organizations. While most of these titles are brazenly prescriptive, some are more descriptive, but they all tend to emphasize the monumental change brought about by digital technology. If we were to believe all the hype about 'digital disruption' and 'technological revolution', we would have to conclude that we are now stuck in some kind of brave new world, where our classical theories of organization have become obsolete. Instead of talking about organized production, this new discourse speaks of 'produsage' on digital platforms; and instead of looking for the power dynamics within organizations, the buzzword is 'empowerment through digital technology'. In short, there is a hype promoting digitization as an unstoppable force of change, with mostly positive consequences. The organizational dimension, if mentioned at all, is practically little more than an afterthought.

The present book was born as a reaction to this new wave of celebrationist digitalization literature. We first discovered the surprising lack of *organizational* accounts of digital technology five years ago, while teaching a graduate course in 'Digital Organization' at Copenhagen Business School (CBS). Having to construct our syllabus from scratch, we found ourselves struggling to find the kind of textbooks that could truly unpack the organizational repercussions of digital technology's omnipresent nature in contemporary society. All too often, we had to draw on literature from other disciplines such as sociology and media studies, and then tweak these into dialogue with our own field of organization studies. However, this strategy ended up being unfulfilling. It ignored the fact that technology has always been with us in our effort to understand how organizations work. In fact, technology lies at the heart of some of the most canonical texts within organization and management studies such as Joan Woodward's work on industrial organizations and Charles Perrow's studies on complex organizations. How come this vibrant scholarly interest in technology and organization has not made itself visible in textbooks on digitalization? Our goal with this book is to offer students an organizational perspective on digitalization.

Although we could have taught our course by relying on scholarly articles on digital technology and organization, we decided to embark on a more ambitious project: writing a textbook that could unpack several aspects of digital organizing, a textbook that could be used not only in our own courses at CBS, but in other universities and business schools as well.

PREFACE

As such, our goal with this book is to provide an organizational vocabulary as an alternative to the hyped-up narratives of digital transformations. We do this by revisiting some of the classical themes in organization studies, in order to show how digital technologies have become intertwined with different aspects of organizational life. The structure of the book reflects this ambition. The first three chapters build a foundation for talking about digital organizing as a specific practice. We then provide seven chapters, each focusing on a specific theme relevant to organizational practices: structure, production, collaboration, knowledge, communication, legitimacy, and power. In the final chapter, we discuss the implications of digital organizing for individuals, organizations, and societies as a whole. Throughout the book, we retain our focus on digitalization from an explicitly organizational point of view. Instead of having digital technologies as the starting point when we talk about organizations, we reverse the equation: our goal is to talk organizationally about the digital. The entrance of new digital technology into organizations, we argue, is not just a technical issue, but also an organizational challenge in many ways. If we can make the organizational aspects of digitalization clearer to the reader, then we will have succeeded in our mission.

When we began writing this textbook, one of our colleagues teasingly remarked that our themes were 'so 1980s'. It was as if concepts such as structure, production, collaboration, knowledge, communication, legitimacy, and power were somehow out of date. Why were we not crafting our book around more topical themes like co-creation or disruption? This subtle challenge pushed us to reflect on the relationship between the classical themes in organization studies and more contemporary concepts. Indeed, we decided to face up to the challenge of re-actualizing the classical themes and to show how they are still essential to understanding organizations in the digital age. We therefore argue that the classical problems of organizing do not disappear with the spread of digital technologies. With our ambition to demonstrate the continued utility of the classical organizational theory concepts, even while acknowledging the value of newer approaches, writing this book was far from easy. Beginning each new chapter was like entering two competing terrains of knowledge and expertise. Embarking on a quest to examine 14 corners of academic literature – seven classical and seven more contemporary themes – proved to be quite a daunting task. While we obviously could have discussed certain issues in more depth, we have tried to provide enough conceptual resources to allow students to speak organizationally about important aspects of digitalization.

A number of colleagues have helped us complete this book project, and they all deserve recognition and appreciation. The idea of speaking organizationally about important aspects of digitalization is shaped by a particular academic environment – the Department of Organization at CBS. In writing this book, we have also been influenced by the ongoing research project on the transformation of work in digitalized public sector organizations that

PREFACE

Ursula has been carrying out with our colleague, Lise Justesen. While the results of their project have been published in academic papers, we have also sneaked empirical examples from their research into some of this book's chapters as illustrations. Similarly, ideas developed through numerous discussions and shared observations have helped refine the way we conceptualize the practice of digital organizing as described in this book.

Apart from this inspiration, our greatest thanks go to our dear friend, Andreas Kamstrup, who was once part of the project, but who cleverly realized that life outside academia offers many benefits over life on the inside. Andreas helped us conceptualize the book and offered his help whenever needed. In addition to Andreas, we also want to thank those colleagues who generously read and commented on various chapters: Mikkel Flyverbom, Signe Vikkelsø, Morten Knudsen, Liana Razmerita, Anne Roelsgaard Obling, Hans Krause Hansen, Elena Raviola, Oana Albu, Cecilie Glerup, Lise Justesen, Peter Holm Jacobsen, and Trine Pallesen. We also wish to thank our editors and editorial assistants at Red Globe Press, as well as the anonymous reviewers, for their valuable advice on our book proposal and the individual chapters. When our themes and arguments were more or less in place, Mette Plesner helped us visualize them through creative image editing and graphic illustrations. The book would not be the same without her! Last but certainly not least, we would like to thank our families, Mikkel and Tanja, our partners in crime, and our children – Lucca, Theo, Karla, and Luis – simply for being there and for being them. With all our hearts, thank you!

Ursula and Emil
Copenhagen, September 2019

PART I

DIGITAL ORGANIZING

1 THE DIGITAL AND THE ORGANIZATIONAL

This book is titled *Digital Organizing*, and the present chapter zooms in on the implications of these two words. We begin our account of the digital in the 1850s, when we saw the first attempts to develop computational machinery and the binary thinking that forms the basis of present-day digital technologies and the automation of work. We then jump to the 1940s, when advances in engineering resulted in the transistor, which allowed computational power to double every 18 months since then. This exponential growth in computing capacity has resulted in an enormous amount of innovation within, and diffusion of, digital technologies and their applications in organizational contexts. We end our account of the digital with a discussion of how a new epoch of connectivity between individuals and organizations began in the 1960s, when the first steps toward the invention of the internet were taken. The first part of this chapter is therefore dedicated to fleshing out what is implied by the terms 'algorithmic thinking', 'exponentiality', and 'connectivity'. These terms are important in understanding the technical foundations of digitalization as we experience it today. Based on this very basic understanding of what the term 'digital' refers to, we can then discuss what is implied when we talk about digitalizing products, processes, and entire organizations or industries.

Following the description of 'the digital', the second part of the chapter presents our approach to thinking organizationally about digitalization. We explain how organizational thinking differs from more technology-oriented perspectives on digitalization. Our goal here is to show that although new organizational dynamics emerge with digitalization, the established concepts and approaches from organization studies remain useful tools for understanding the digitalization of organizations. The chapter ends with a presentation of the structure of the book, describing seven major organizational themes: structure, production, collaboration, knowledge, communication, legitimacy, and power. We will show how each of these classical concepts can be linked up with a more contemporary concept, which together can elucidate various dimensions of digital organizing not necessarily captured by the classical concepts in their original form.

Algorithmic thinking

The idea of producing machines that could compute arithmetical tables dates back to the early nineteenth century. Sadie Plant tells the poetic story of how, in 1833, an amazed audience was invited to inspect a futuristic device "which seemed to have dropped into [the] world at least a century before its time" (Plant, 2016, p. 5). The device in question was the Difference Engine, invented by the engineer Charles Babbage. Babbage's project was to use machines to compute arithmetical tables, and his Difference Engine was able to raise "several Nos. to the 2nd and 3rd powers, and extracted the root of a quadratic Equation", as one of the onlookers observed (Plant, 2016, p. 5). That same year, George Boole (1854) developed his 'symbolic logic', later published in his *An Investigation of The Laws of Thought*. Based on formal logic and algebra, the book presents the idea that logical relations can be expressed in symbolic form: thus, the logical dichotomy 'true or false' can be translated into the binary digits '0' or '1'. As Boole himself writes, his objective was "to investigate the fundamental laws of those operations of the mind by which reasoning is performed; to give expression to them in the symbolical language of a Calculus" (Boole, 1854, p. 1).

Today, we talk about Boolean algebra as that branch of algebra where the values of the variables are 'true' and 'false' and the main operations are 'and', 'or', and 'not'. Consider how Boolean algebra differs from elementary algebra, where the values of the variables are numbers and the main operations are addition and multiplication. Boole's work laid the foundation for the design of modern computers. Now that logical propositions could be expressed in algebraic form, logical deductions could be drawn by algebraic calculations. But it would take decades before operational computers could be designed and built. In the 1930s, Alan Turing developed a theoretical model of the computer: "With a tape drive and a computation unit, this hypothetical, abstract machine was capable of reading, erasing, and writing digits on a single line of type. It processed zeroes and ones on a tape of infinite length which passed through the drive, and followed a series of basic commands" (Plant, 2016, p. 82).

As Plant notes, all "subsequent computers are implementations of this most general of general purpose machines" (Plant, 2016, p. 82). Early computer technology (Image 1.1) was developed by engineers working with electronic circuits. An electronic circuit can be in an 'on' mode and an 'off' mode, corresponding to a '1' when the current passes through the circuit and a '0' when the current does not pass through it. This again corresponds to the Boolean 'true' and 'false'. The basic operation of a computer starts with these 'on' and 'off' circuits, which are interpreted as '1' or '0'. The ones and the zeroes are ordered into patterns, and the patterns are what gives them meaning. To take a random example, the pattern 1110001100 corresponds to the number 908. However, it is not only numbers that can be represented by zeroes and ones. Any object or experience that is produced or processed by digital technologies (graphics, sound, photo)

THE DIGITAL AND THE ORGANIZATIONAL

Image 1.1 Computer scientist Grace Hopper presenting a computer in 1952. Hopper popularized the idea of machine-independent programming languages, and she is credited with popularizing the term 'debugging' for fixing computer glitches – inspired by an actual moth removed from the computer.

Source: Science Source, Ritzau/Scanpix.

consists of combinations of zeroes and ones. Computer operations are also just numbers, encoded to make different things happen. Even the most advanced software is nothing more than a very long set of on and off signals, which – combined – tells a computer what to do (Image 1.2). The sets of rules defining the operations that a computer must perform are called algorithms. Since we are interested in the organizational dimensions of digitization, we will not delve further into a discussion of binary logics or computation as such. We will simply underscore here that binary logics are the foundation for the development of both algorithms and computers and argue that they are also becoming an important force in structuring organizational life.

Binary logics in organizations: The case of digitization-ready legislation

The binary logics discussed here, besides being foundational to computer programming, have also become a factor in redesigning organizations, ostensibly to make them more efficient. The goal is that certain types of routine administrative tasks can be carried out by algorithms that can process information and even 'make decisions'. Let us take an example from the organization of public service delivery. In countries where the digital infrastructure is advanced, public organizations that deal with certain kinds of standard administrative case processing can eliminate a number of meetings between citizens and public servants, and all the interaction previously required in connection with preliminary registration, application forms, information storage, processing, and decisions that needed to be made in relation to a case. Some cases may be quite simple: in a country that pays child subsidies, it is enough that the 'system' is informed that a woman has given birth to a child. Based on the mother's national ID number, the system then accesses the parents' income, determines the amount of payment, and sets up a monthly transfer of child subsidy to the mother's bank account. This happens in Denmark, where the public sector is among the most digitized in the world, and where all persons have a national ID number and a linked bank account. In the Danish public sector, much of the routine administrative work has become automated in this way (Justesen & Plesner, 2018). Data about the citizens are already available in digital platforms shared across the public sector, so when citizens apply for various services online by entering specific information, their application can in some cases be processed by an algorithm, untouched by human hands, or unread by a human case officer. An algorithm is programmed to produce a 'decision' and then inform the citizen immediately that they are accepted/rejected for a subsidy, or that they are being refunded X amount of income tax.

The efficient use of algorithmic decision-making requires that the law and the procedures be simple. If this is the case, an algorithm can quickly decide whether a citizen is entitled to a particular social benefit on the basis of objective factors such as age or income level. To make the most of this technological possibility – and to achieve the potential benefits in efficiency – all new legislation in Denmark needs to be 'digitization-ready'. This means that the law – to the extent possible – must be written in simple, unambiguous terms that can be translated into binary codes for use in the

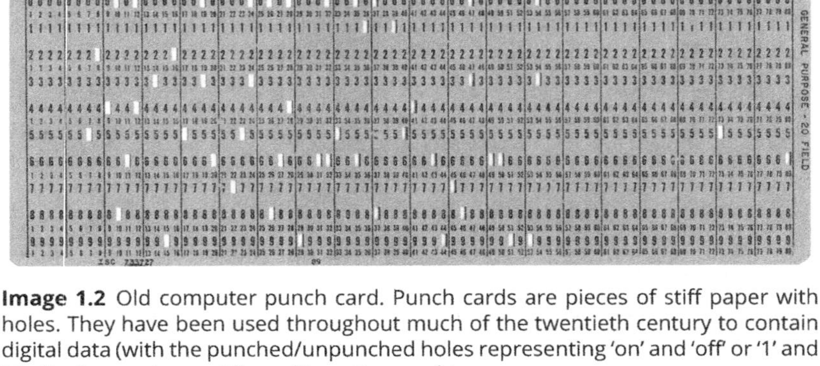

Image 1.2 Old computer punch card. Punch cards are pieces of stiff paper with holes. They have been used throughout much of the twentieth century to contain digital data (with the punched/unpunched holes representing 'on' and 'off' or '1' and '0'). Similar cards are still used in voting machines.

Source: EyeJoy, iStock.

administration of the law. The automation of work processes presupposes binary options and a consistent vocabulary. This is because even the most sophisticated algorithms cannot handle ambiguous categories or problems. Complex administrative laws call for professional judgment and can end up preventing the use of digital solutions. Only a human employee can consider special cases or interpret ambiguous language (Justesen & Plesner, 2018). In the Danish public sector, the automation of administrative work based on binary thinking has led to extensive analyses of work processes in order to determine what kind of work can be automated and what needs to be handled by people.

The digitalization of case-handling raises numerous questions: How can we ensure that automation enhances the quality of services? What does it mean to be treated 'fairly'? What role should programmers play in the legislative process? And, most importantly, are there limits to automation? Are there certain processes that are so complex or sensitive that they should require face-to-face interactions and human judgment? Given the speed of technological advances and the constant quest to improve efficiency, digitalization is becoming more a question of *how* we follow binary logics to automate organizational processes rather than *whether* we should automate certain processes.

The phenomenon we have described here is also known as computational thinking. Computational thinking attempts to re-define a problem so that it can be solved by a computer. This process usually entails breaking down a complex problem into ever smaller, more manageable parts (Google for Education, 2017). This way of thinking is both essential to creating computer applications and creates the basis for the digitalization of organizations (Table 1.1). Across many types of organizations, computational thinking is applied in analyzing work processes, simplifying decision-making, and automating case-handling.

Table 1.1 The concepts of digitization and digitalization.

	Definition	Example
Digitization	The process of translating *information* from analogue to digital formats	The digitization of literature, as carried out by the Google Books project
Digitalization	The use of digital technology to change organizational processes and practices	The digitalization of the taxi industry, as carried out by Uber or Zoro

Beyond binary thinking: The challenge of complicated mathematics

'Digitization-ready legislation' and the resulting reconfiguration of public administration are examples of digitalization of organizations based on algorithmic thinking. The examples emphasize the practices of extracting the human element from the equation so that work processes can be automated. But this simplification process is not the full story of algorithmic thinking. Although algorithms are based on simple binary logics, they can perform very complex tasks. They can be designed to organize decision-making, services, trade, and entire industries in ways that are impenetrable for anyone who is not a skilled mathematician. This is the point made by Cathy O'Neil (2016) in her book *Weapons of Math Destruction*. A mathematician herself, with a PhD in algebraic number theory, O'Neil worked for a hedge fund, joined a risk analysis company, moved to an internet start-up, and finally decided to use her insights about mathematical models to enrich the public debate. Based on her experiences in the field, O'Neil has grown increasingly skeptical of how we allow algorithms to make quick, but ultimately opaque, decisions:

> This was the Big Data economy, and it promised spectacular gains. A computer program could speed through thousands of résumés or loan applications in a second or two and sort them into neat lists, with the most promising candidates on top. This not only saved time but also was marketed as fair and objective. After all, it didn't involve prejudiced humans digging through reams of paper, just machines processing cold numbers. By 2010 or so, mathematics was asserting itself as never before in human affairs, and the public largely welcomed it. Yet I saw trouble. The math-powered applications powering the data economy were based on choices made by fallible human beings. Some of these choices were no doubt made with the best intentions. Nevertheless, many of these models encoded human prejudice, misunderstandings, and bias into the software systems that increasingly managed our lives. Like gods, these mathematical models were opaque, their workings invisible to all but the highest priests in their domain: mathematicians and computer scientists. (O'Neil, 2016, p. 3)

In her book, O'Neil discusses a number of areas where organizations rely on advanced algorithms to process data and to make decisions that may end up having grave consequences. A spectacular example is that of a very popular schoolteacher who was fired because of a score generated by a 'value-added modeling system'. The school district had hired a consulting firm to develop an evaluation system that could rate teachers in order to weed out the worst of them (those in the bottom 2 percent). The algorithm used was supposed to eliminate human bias and be objective. At the same time, it was bound to be very complex because it needed to integrate a myriad of factors such as students' socioeconomic background, learning disabilities, various measures of educational progress, teacher evaluations, and so on. In the end, this meant that, although the teacher got excellent reviews from her principal and from her pupils' parents, and herself believed that she was a good teacher, she got laid off due to the low scores. Moreover, she could never obtain any explanation for how these scores were derived. In O'Neil's analysis, the first problem with the algorithmic basis for firing the teacher was that nobody could explain how the score had come about. "It's complicated," as the managers said. Some of the more fundamental problems with the algorithm were that it could not judge whether the data entered into the system were valid, and that it was designed to make inferences on the basis of a very low number of cases – which is statistically unsound.

Despite such weaknesses, it is clear that algorithmic thinking has become important in the reorganization of work and the relations between employees, managers, and citizens or customers. It is also worth noting how much the implementation of algorithms ultimately depends on the mathematicians who develop them. Normally, mathematicians and statisticians live a quiet life as number crunchers. Today, they are increasingly sought-after employees, exerting major influence on the way digitalization is spreading across all aspects of work. At the same time, because the design of algorithms is so closely linked to organizational design, an interdisciplinary approach is becoming essential in the reorganization of work. The above examples of digitization-ready legislation and the scoring of teachers reveal the many organizational, legal, mathematical, ethical, and managerial issues at stake. In designing the interplay between the digital and the organizational, many voices need to be heard, and there are many interests at stake.

Exponentiality

Let us now move on to another basic component of the digitalization of organizations. The phenomena we just discussed became possible only because we have moved far beyond rudimentary computer technology. Advanced automation and calculation processes became possible with the invention of the transistor. Developed in the 1940s, transistors revolutionized the calculation capacity of computers. They are made from a material (silicon, giving the name to Silicon Valley) which allows electric current to pass through it at the level of electrons, and which has the capacity to magnify the energy that goes into the transistor before transmitting it again. Electrons are tiny particles of the atom, enabling transistors to contain millions of electrons. Microchips, in turn, contain millions of transistors, and their storage and calculation capacity are what makes all present-day digital devices work the way they do. Transistors have thus been called the 'nerve cells' of our digital age (Riordan & Hoddeson, 1997).

The Silicon Valley firm Intel continues to be the most successful producer of microchips. In 1965, one of its founders, Gordon Moore, predicted that transistors would become cheaper and increasingly efficient, to the degree that computational power would double every year. This prediction, based on early empirical observations, proved accurate for several decades and was termed 'Moore's law'. Moore's law implies that computational power increases exponentially. It is an open question whether this development will continue, as it would demand advances in innovation in materials other than silicon. But a group of people based in Silicon Valley (with associates around the world) believes that new technologies will pave the way for a continuation of the law, such that artificial intelligence will end up surpassing human intelligence. The group consists of futurists, engineers, and others who call themselves Singularity University. They share technology updates and forecasts as well as encouragements to "transform your mindset" and embark on "a transformational journey to the future" (Singularity University, 2018).

To be sure, the increase in computational power already means that machines can outperform humans in a variety of tasks. This is due largely to the ability of computers to sift through, analyze, and act on the basis of enormous amounts of data. Research and experiments in artificial intelligence have the goal of enabling machines to learn by themselves. Machine learning and artificial intelligence are applied on many different levels, and there are grandiose hopes attached to the exponential potential. As Erik Brynjolfsson and Andrew McAfee (2014) note: "Not only are the new technologies exponential, digital, and combinatorial, but most of the gains are still ahead of us. In the next twenty-four months, the planet will add more computer power than it did in all previous history. Over the next twenty-four years, the increase will likely be over a thousand-fold" (p. 32).

In this book, we do not intend to speculate about the potential of technology or the future of organizations in an increasingly digitized world. We only wish to point out that the digitalization of organizations raises important questions that we need to tackle here and now, particularly issues of how work is organized around digital technologies, how a certain transparency in relation to the functioning of complicated mathematical models can be established, and how decisions are delegated to machines.

Delegating decisions to machines: The case of the driverless car

The complexities of predictions become obvious in the case of self-driving cars. As Brynjolfsson and McAfee (2014) observe: "self-driving cars went from being the stuff of science fiction to on-the-road reality in a few short years. Cutting-edge research explaining why they were not coming anytime soon was outpaced by cutting-edge science and engineering that brought them into existence, again in the space of few years" (p. 19).

While Brynjolfsson and McAfee are certainly correct about the rapid advances in digital technologies and artificial intelligence, we have only begun to grapple with the problem of how advanced digital systems can operate beyond human control and understanding. These problems compel us to take a few steps back. Brynjolfsson and McAfee had experienced a successful ride on US Highway 101 in a driverless car at the time when they wrote their book. Some years later, however, a driverless car failed to recognize a pedestrian crossing an intersection, striking and killing her (Smith, 2018). As the car was in an unpredictable situation, it had been programmed to hand over the control to the human driver. However, the human driver was not paying attention and could not intervene quickly enough to avoid the accident. Since then, driverless test cars have been physically attacked by angry citizens, slashing their tires and creating obstacles that hinder their operation. Here, popular sentiments and public opinion, and not unreasonable anxiety about runaway technology, become an essential element in how we need to think of technological advances.

The case of driverless cars raises questions about intelligence and contextual understanding. Machines have become better at adjusting to their contexts and rewriting their own codes. However, these advances produce a well-known set of hazards, similar to complicated stock-trading algorithms that mechanically produce panic sell-offs when they take over trading. The unintended consequences of algorithmic decision-making become increasingly important to consider as software code becomes more complex. The concept of 'spaghetti code' is used to describe a mess of algorithms that interact in unpredictable ways and generate anomalous, undesired output (Smith, 2018). The mundane, organizational tasks required to adapt to these technological developments include those related to regulation, responsibility, and ethics. It is clear that programming is not just a technological activity. Tech firms and tech innovators have an enormous responsibility in relation to the wider effects of their work. This extends to organizations that embark on digitalization projects. We need to consider the broader implications of projects that might seem to be benign or innocent on first sight. This kind of reflection lies at the core of dealing responsibly with digital transformations.

Connectivity

We have discussed algorithmic thinking and exponentiality as two technological aspects of the digitalization of organizations. A third aspect is connectivity, or the increasing interconnectedness of computers. Moving information from analogue to digital formats is digitization in a very elementary sense. Widespread, pervasive digitalization of interactions depends on the technological infrastructure of the internet, or connectivity (Image 1.3).

The first steps toward the invention of the internet as we know it were taken in the 1960s, when the US Department of Defense funded the creation of the Advanced Research Projects Agency Network, or ARPANET. The ambition was to connect isolated computers so that data could be shared more easily. ARPANET was the first successful network of computers in which data could be transmitted in small packages.

Image 1.3 Connectivity is a matter of linking nodes into a network.

Source: Gremlin, iStock.

Up through the 1980s, it remained technically very difficult to access all this information, even if computers were connected. In 1989, however, software engineer Tim Berners-Lee developed the idea of a 'distributed hypertext system', which eventually became the foundation for the World Wide Web (CERN, 2019). Berners-Lee developed the fundamental technologies of the internet as we know it today (HTML, URL, and HTTP). HTML stands for HyperText Markup Language, which is the formatting language for the web. URL stands for Uniform Resource Locator and functions as the digital address used to identify each resource on the web. Finally, HTTP stands for HyperText Transfer Protocol, which allows for the retrieval of linked resources from across the web. Berners-Lee wanted the code and standards of the internet to be free and accessible, and this became the basis for the bottom-up creation of the internet in all its complexity – and for the open-source movement, which we will discuss below.

The ethos of openness

The ethos surrounding the internet in the early days was very much about decentralization, non-discrimination, bottom-up design, and universality (World Wide Web Foundation, 2019). The internet has no central control, it has equal access, its code is free to use and develop, and it offers the same possibilities of expression to everyone. This ethos, including the idea of net neutrality, has recently been severely challenged by, for instance, government censorship and large tech companies that have succeeded in capturing users' data and then marketing them. Hence, while connectivity in the early days of the internet may have stimulated new approaches to information (Open Data), politics (Open Government), and scientific research (Open Access), and the spirit of sharing, connectivity has also become the foundation for a new economy based on the commercial exploitation of shared data and networked products and services (see e.g. Galloway, 2018; Zuboff, 2019).

In attempts to capitalize on the original idea of 'no central control', the encrypted transaction technology blockchain has been developed to make transactions easier and more controllable by users rather than intermediaries. Blockchain technology has been developed for both financial transactions and other business purposes, and pushed by players like IBM. In their book *Blockchain for Dummies*, they describe how and why:

> The blockchain architecture gives participants the ability to share a ledger that's updated through peer-to-peer replication each time a transaction occurs. Peer-to-peer replication means that each participant (also called a node) in the network acts as both a publisher and a subscriber. Each node can receive or send transactions to other nodes, and the data is synchronized across the network as it's transferred. The blockchain network is economical and efficient because it eliminates duplication of effort and reduces the need for intermediaries. It's also less vulnerable because it uses consensus models to validate information. Transactions are secure, authenticated, and verifiable. The participants in both transactions systems are the same. What has changed is that the transaction record is now shared and available to all parties. (Gupta, 2018, p. 6)

As with the basic internet technology, the invention of blockchain has a dual face. It is hyped for its liberating potential because it offers secure transactions, and can therefore be used as yet another tool for making money. In a kind of utopian manifesto, blockchain is proclaimed as the weapon that will undermine organizational control:

> With Blockchain, we can imagine a world in which contracts are embedded in digital code and stored in transparent, shared databases, where they are protected from deletion, tampering, and revision. In this world, every agreement, every process, every task, and every payment would have a digital record and signature that could be identified, validated, stored, and shared. Intermediaries like lawyers, brokers, and bankers might no longer be necessary. Individuals, organizations, machines, and algorithms would freely transact and interact with one another with little friction. This is the immense potential of Blockchain. (Iansiti & Lakhani, 2017, p. 120)

At the same time, more cautious voices observe that blockchain "will presumably not eradicate the organization but could change the understanding of it and may lead to a different answer about the nature of the firm" (Scholz & Stein, 2018, p. 2).

Cases of blockchain technology in practice

Proponents of blockchain technology cite its ability to solve problems of corruption, ensure ownership, and facilitate tracing. To take an example, the company Provenance is dedicated to helping businesses build trust in their goods and supply chain. As stated on their homepage: "Powered by mobile, blockchain and open data, our game-changing software enables retailers and producers to open product data, track the journey of goods, and empower customers with access to knowledge" (Provenance, 2018). The company has developed a platform where all the participants in a transaction have access to the same data – it is impossible for one partner to control the transaction and impossible to delete information because it is locked and decentralized. This kind of system can create transparency in food supply, for instance, where data about the source, transportation, and packaging can become accessible to consumers on a mobile device. Blockchain has its dark sides, of course: the encrypted currencies afforded by blockchain technologies can facilitate illegal transactions, hacking, cyber blackmail, or evading established institutions because secure transactions can take place between individuals, without intermediaries. What sometimes seems like elusive, online phenomena actually affects material and financial realities.

Connectivity as a material issue

Following this line of thinking, we should also remember that connectivity through digital technologies operates not only on the ideational or symbolic level, but also on the material level. The Internet of Things is the expression for connecting objects and sensors to the internet so that they can provide various types of information that can be automatically processed. Connectivity of things allows for the automation of another kind of work than the administrative processes described above. Let us provide a few examples. Through the Internet of Things, trash cans can transmit signals when they are full, and algorithms can design routes for trash collection based on these data. Printer cartridges can transmit signals when they are nearly empty, and then place an order to a supplier to purchase and ship a replacement cartridge, all automatically.

The calculation power and storage capacity needed to deal with all these data comes from enormous data centers. Hence, when we speak about saving data 'in the cloud', this is a euphemism for storing them in a network of computers rather than on our own hardware. The data centers have nothing to do with clouds. They are right here on earth, require extraordinary amounts of electricity to run, and they emit CO_2 to a degree where they compete with air traffic. Data is not altogether a flimsy, immaterial resource. It requires adequate storage, sophisticated transmission technologies, updated annotations, continuous filtering, and vigilant 'cleansing'. All these dimensions of data put pressure on organizational resources, but curiously, they are relatively overlooked (see also Chapter 5). Connectivity also depends on extensive technical work, but the celebration of technology tends to drown out the persistent technical failures or breakdowns that affect all of us in our work and free time (Graham & Thrift, 2007) and have sometimes dramatic implications for organizational life. When new technologies are introduced into organizations, the expectation is invariably that of increased efficiency and cost reduction. There is much less willingness to assess, much less to calculate, the cost of introducing the technology, the predictable technical failures, or the expected breakdowns that any new system brings with it (in fact, it is mostly with a major catastrophe – a space shuttle explosion, a train wreck, a power outage – that the downside of technology becomes a subject of public discussion).

Digitalization and/or disruption?

In this chapter, we have highlighted algorithmic thinking, exponentiality, and connectivity as central to the digitalization of organizations (Figure 1.1). Among many other technological components and developments, we cite these three because they are foundational to the invention of new products, organizational processes, and practices.

It is a truism to declare that digital technologies have rapidly transformed how we communicate, trade, shop, coordinate, produce, and manage. More and more products and processes are 'made digitizable' – that is,

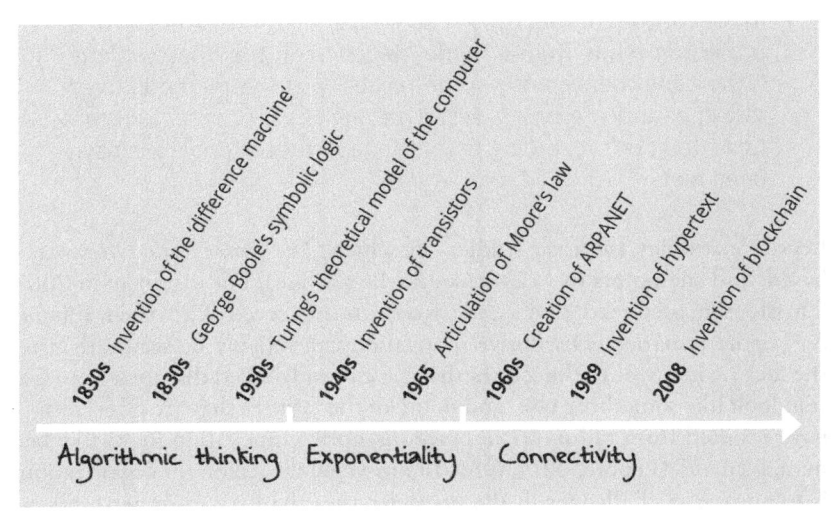

Figure 1.1 Elements in the historical development of 'the digital'.

redesigned on the basis of binary codes, operating on an ever-more refined digital infrastructure, and capitalizing on access to massive amounts of data shared by individuals and organizations. Digital technologies have obviously also transformed how people experience themselves, their relations, and the world around them, and they have made it possible to act differently upon the world. With these developments in mind, it is not surprising that digitalization is often described in grand, future-oriented narratives both in the public debate and in popularized research. Some of the most enthusiastic proponents "celebrate it as a means to finally solve efficiency and quality problems". Others warn us about "the rise of the robots" (Ford, 2015) or the massive unemployment that will follow in the wake of Artificial Intelligence (Susskind & Susskind, 2015)" (Plesner et al., 2018, p. 1178). The sheer fact that computing power has grown exponentially since the mid-twentieth century has led to the belief that we need to brace ourselves for a complete disruption of our economy, our organizations, and our personal lives. Terms like 'Industry 4.0' or 'a new industrial revolution' are coined to capture this kind of thinking. Since the publication of Clayton Christensen's book, *The Innovator's Dilemma* (1997), the terms 'disruptive innovations' and 'disruption' have become ever more widely used. We now have disruption consultants, disruption conferences, disruption degrees, and government disruption task forces. The rhetoric of disruption meets very little criticism, probably because predictions about disruption showcase obvious examples of how businesses or industries already operate differently in the digital age. But as Jill Lepore (2014, n.p.) puts it in a critical piece in *The New Yorker*:

> Disruptive innovation as a theory of change is meant to serve both as a chronicle of the past (this has happened) and as a mode for the future (it will keep happening). The strength of a prediction made from a model depends on the quality of the historical evidence and on the reliability of the methods used to gather and interpret it. Historical analysis proceeds from certain conditions regarding proof. None of these conditions have been met.

Lepore examines the case studies on which *The Innovator's Dilemma* is based, and she points to weaknesses in the analyses. For instance, in 2007, Christensen predicted that Apple would not succeed with their iPhone. As Lepore remarks: "Disruptive innovation can reliably be seen only after the fact ... many of the successes that have been labelled disruptive innovation look like something else, and many of the failures that are often seen to have resulted from failing to embrace disruptive innovation looks like bad management" (Lepore, 2014, n.p.). In our view, the lesson for organizations is to engage in challenges in the present rather than try to understand the 'natural laws' of disruption.

Thinking organizationally about digitalization

Predictions about the near or distant future, based on either historical or present-day cases, might be less informative than close-up studies of how organizations actually change with the development and uses of new digital technologies. Neither utopian nor dystopian accounts tell us much about how digitalization changes work practices at the everyday level (Brown et al., 2017; Plesner et al., 2018). "Whereas the dystopian accounts focus mostly on job redundancies, evidence suggests that new tasks, structures, roles, relations, and responsibilities follow from digitalization. The sweeping narratives often ignore these multiple consequences of digitization (Wajcman, 2017)" (Plesner et al., 2018, p. 1178). In the following section, we discuss what it might mean to think 'organizationally' about digitalization rather than 'digitally' about organization. We do so by discussing two concepts – *bounded* automation and *blended* automation – that can give us a more sober approach to digital organizing than the sweeping narratives of disruption.

Bounded automation

Intervening in the debate about the future of work and organizations, Peter Fleming points to the parallel between the disruption discourse accompanying digitalization and the reactions to the first industrial revolution. In the 1930s, John Maynard Keynes predicted that machines would abolish work within two generations, and computerization led to the same kinds of predictions in the 1980s and 1990s. According to Fleming, however, work seems to be thriving in the artificial intelligence (AI) and robotics era because "machines are not the real issue here. Organizations are" (Fleming, 2019, p. 24). He continues: "Technological innovations do not simply unfurl according to their own endogenous potential. They are delimited by socio-organizational forces, which regulate why, how and whether a job or task is automated" (Fleming, 2019, p. 24).

Fleming calls this phenomenon 'bounded automation'. In arguing for the idea of bounded automation, Fleming takes his point of departure in the disruption discourse described above, citing consultancy reports about the future of work. The McKinsey Global Institute and the Oxford Martin School have produced large-scale analyses of a range of occupations and roles, estimating the percentage of jobs that can be automated. The claim – widely reported by the media – is that the vast majority of routine and semi-routine jobs, which make up about 50 percent of the US economy, can be carried out by robots. Even more dramatically, a New York Times reporter noted that the Davos meeting in 2019 was all about racing to automate workforces: Earlier, people "had incremental, 5 to 10 percent goals in reducing their work force. Now they're saying, 'Why can't we do it with 1 percent of the people we have?'" (Roose, 2019).

However, Fleming emphasizes that the *potential* to replace jobs is not the same as its empirical realization: "Many of the dramatic discussions about robotics and digitization ... often jump from the self-contained achievability of certain innovations ... to their broader, systematic organizational use" (Fleming, 2019, p. 27). This organizational use depends on socioeconomic factors such as the cost of labor, organizational power relations, and the nature of the work task itself. When labor is cheap, it is often not worthwhile automating it. On the other hand, if the labor force is unstable, the option of automating may provide benefits to a given business. Finally, some tasks are widely believed to demand a living person to exercise judgment and induce trust (Image 1.4). Fleming points to certain job categories that might be less susceptible than others to automation. For instance, elite workers will often possess technological expertise, managerial responsibilities, and links to power elites, putting them in a privileged position to manage or oversee the digitalization of organizations. Occupations that can be semi-automated still allow room for employees who can work with or around technology. Finally, there are a number of jobs that are simply not worth automating because labor is cheap and a technological solution would be more expensive.

Fleming's concept of bounded automation points to all the organizational factors that play a part in constraining automation. In this way, "[v]iewing automation in 'the second machine age' as an organizationally delimited phenomenon permits us to shift attention away from technology per se" (Fleming, 2019, p. 31). Hence, digitalization certainly has an impact

Image 1.4 A robot priest wearing a Buddhist robe stands in front of a funeral altar during a demonstration. As Fleming (2019) writes, just because there is a *potential* to automate jobs, automation is not necessarily carried out. The priest might be a good example of a job function that we prefer to keep being done by humans.

Source: Kim Kyung-Hoon, Ritzau/Scanpix.

on work and organization, but this impact needs to be considered *in organizational terms*. Digitalization is an *organizational* process.

Blended automation

Fleming's idea of 'bounded automation' is confirmed by organizational research, both that concerned with the introduction of automation and that which discusses rolling back automation. For example, Daniel Beunza studied how the social transactions of trading on the floor of the New York Stock Exchange (NYSE) from 2003 to 2012 were replaced by automated trading. However, it turned out that the fully automated trading floor could not cope with situations of crisis, so a new model was introduced. Daniel Beunza and Yuval Millo (2015) call this 'blended automation'. They observe that although trading algorithms are highly efficient, they were unable to deal effectively with environmental complexity. These complex situations require people who interact in a social system, who can make sense of ambiguous news, anxious rumors, and unexpected events, and who can enforce social norms.

Blended automation can be seen as a specific type of organization design, in which the most routine activities are automated while humans intervene when something extraordinary takes place or when adjustments to environmental complexities are needed. As Beunza and Millo write:

> there are situations when market complexity or opportunism calls for the skilled activities of the manual actors. Blended automation addresses this problem by preserving the skilled actors and their routines and infrastructure (duplication), allowing them to remain available to address exceptional situations (exceptionality). The combined effect of duplication and exceptionality can readily be observed in the case of airplanes and autopilots, which rely on a combination of duplication (autopilot and the pilot) and exceptionality (reliance on the pilot for take-offs, landings and turbulence) akin to the Exchange. As with the NYSE, the autopilot runs on a simplified representation of the navigational environment that is useful for most situations, but falls short in cases of high environmental complexity; in those, the presence of the pilots allows for their skills to be brought to bear. (Beunza & Millo, 2015, p. 37)

The concepts of bounded automation and blended automation serve to remind us that despite the spectacular developments in digital technologies, these technologies and digitalized phenomena generally do not operate in a void. They are driven by people and often embedded in organizations (Brown et al., 2017). Taking an explicitly organizational perspective on digitalization compels us to consider some of the issues that are interwoven with digital technologies in practice. Moreover, an organizational approach can restrain us from jumping on the future-telling bandwagon and reproducing grand narratives of disruption. In fact, for many industries or organizations, it is still an open question whether they will be 'disrupted' or whether they will simply be reorganized around new digital technologies (Image 1.5).

It is a well-established tenet of organization studies that technologies and organizations are 'intertwined', or that they mutually affect one another. It thus seems fair to assume that when new technologies are introduced into an organization, this will lead to new work tasks, functions, and professional roles and relations. When thinking organizationally (Du Gay & Vikkelsø, 2017; Kamstrup, 2017) about digitalization, it is important to emphasize that new practices or roles do not emerge in any simple or predictable way. Instead, it is an empirical question how such changes related to digitalization unfold, and an analysis needs to take the role of organizing into account. As Bruno Latour (2013) notes, if we do not speak organizationally about such matters, it is "as if you wanted nature to speak directly without the institutions of science or flowers being delivered directly through the wifi" (Latour, 2013, p. 170). The point is that organizing matters. In Latour's examples, the organizing performed by scientific institutions matters and the organizing performed by florists matters to how science develops or

Image 1.5 A New York Stock Exchange screen showing the closing stock price results. Organizations across the financial sector are wrestling with how to work with and around new digital technologies, and particularly the problem of balancing the benefits of automation with the need for human interventions. Both automation and human interpretations play an important role in financial trading.

Source: Xinhua/Eyevine, Ritzau/Scanpix.

flowers are delivered. The robots *are* coming (as the saying goes), but their impact on contemporary society is molded and conditioned by organizing processes. Our goal in this book is to offer a nuanced understanding of how digitalization can be understood in relation to a range of organizational phenomena, and to offer an organizational language with which to talk about digitalization.

The promise of this book

In pursuing this goal, we provide a systematic discussion of a number of organizational phenomena that can either influence, or be influenced by, the introduction of digital technologies. We focus on some of the central themes of the organization studies literature, and then discuss these themes as they pertain to the digitalization of organizations. Potentially, any organizational phenomenon could be examined in relation to digital technologies. In this book, we have chosen to start by looking into organizational structures – a classical unit of analysis in organization studies. However, our examination of organizational phenomena will not be limited to the formal organization alone, as this would make the present book a book about 'the digital organization'. We also want to revisit a range of issues discussed in other areas of organization studies, namely processes of organizing related to production, collaboration, knowledge, communication, legitimacy, and power.

These processes have become increasingly important objects of analysis in organization studies as the discipline has evolved over the years (Clegg, 2014). This development has happened in conversation with sociology, psychology, political science, and other disciplines, with a focus on how organizing is performed through linguistic and material means. We interrogate a broad range of themes because we believe that to understand organizations and organizing, it is not sufficient to limit ourselves to the purely formal properties of organizations such as the division of labor or hierarchical structures. We also need to understand dynamics between the individuals and groups inhabiting the organization. In addition, we need to understand how digital technologies influence these dynamics between individuals, groups, and organizations. *Both* issues related to organizational structures *and* dynamics relating to production, collaboration, knowledge, communication, legitimacy, and power become increasingly entangled with digital technologies. While these themes have taken on increasing importance in organization studies, the implications of digital technologies for how we understand them have not been systematically addressed. Our book is an attempt to rectify this shortcoming.

Classical and contemporary terms

To underscore that digital technologies actually produce new organizational practices and phenomena, one might be tempted to replace the classical organizational terminology with more contemporary terms connoting the changes brought about by digitalization. Terms such as 'interactivity' or 'co-creation' have become more widespread in roughly the same time frame as the advent of digital technologies. All the classical terms have a more contemporary – and perhaps complementary – counterpart, which we will apply in order to broaden our understanding of digital organizing. Hence, we can speak of 'infrastructure' as an often-overlooked aspect of structure. We can discuss 'produsage' as a new aspect of the conventional concept of production. We can cite 'datafication' as a new phenomenon related to the larger issue of knowledge. We may talk about 'co-creation' as a new set of practices under the heading of collaboration. The concept of 'interactivity' may add new insights regarding communication, 'transparency' sheds new light on legitimacy issues, and 'empowerment' could be seen as an increasingly important aspect of power.

Like any tool or method, the choice of terminology is not neutral, and does not lead to neutral depictions (Law, 2004). Instead, the choice of analytical term is an activity that shapes the object it describes, so an awareness of the implications of this choice is important. We do not want to talk about a distinct 'before' and 'now' in our analysis of organizational themes. We strongly believe that the classical themes and terms in organization studies remain relevant. However, they can be complemented with more contemporary terms that foreground developments in various practices and demonstrate how our language evolves with changing practices. In particular, the contemporary terms we shall discuss throughout this book point to more active technologies, organizational subjects, and technology users. Hence, we maintain that it is untenable to *replace* classical themes, terms, and concerns with new ones. After all, when we deal with developments in digital technologies, we find that the topical quickly becomes outdated. We bring together the more classical and the more contemporary terms because we think, by looking at them as complementary pairs, we can better understand the dynamics and implications of digital organizing. Rather than describing a scenario of 'before' and a 'now', or an 'either/or', we are highlighting the 'both/and' nature of classical and modern terms.

The chapters

This book is situated within the field of organization studies. Hence, the following chapter – Chapter 2 – offers a historical account of how technology has been theorized and studied by organizational theorists. In this chapter, we link key developments in organization theory to the study of different types of technology. The field of organization studies was originally conceived as a technical science closely related to the science of engineering. However, organization studies gradually lost sight of technology. This book is part of a revival of the question of technology in the light of the digitalization of organizations. The chapter proposes a dual focus on the material and the social aspects of organization, and we argue that this dual focus offers a productive perspective on digital organizing.

Like Chapter 2, Chapter 3 can be read as foundational to the rest of the book. The chapter discusses different approaches to technology and organization. We have chosen to provide an overview of what we consider to be the three most prominent approaches to this relationship: technological determinism, social constructivism, and sociomateriality. We begin by considering the idea that technology determines the social, and, by implication, pushes organizations in particular directions. This techno-determinist perspective has two normative subsets – technological pessimism and optimism. An awareness of these two positions can be helpful because they resurface in much of the literature on the digitalization of organizations. We then proceed to discuss the second perspective, that of social constructivism, which emphasizes the social factors that play a crucial role in shaping technology. Finally, we discuss how the sociomaterial perspective tries to strike a balance between technological determinism and social constructivism. Toward the end of the chapter, we introduce the concept of 'affordances'. We argue that because the affordance concept focuses on both the properties and uses of technology, it provides a fruitful way of analyzing digital organizing, regardless of what perspective one employs.

Chapter 4, the book's first thematic chapter, highlights how organizational structures can be seen as central to understanding how organizations manage collaboration, communication, and control to achieve a common goal. Organizational structures can be both formal and informal, they can be the object of redesign, and they can be depicted in diagrams to illustrate connections and division of labor. Organizational structures continue to be a key practical concern, and while they have been a prominent focus of organizational studies, it has also been argued that they tend to become static and out of sync with more dynamic and messy organizational realities. Such critiques are discussed in this chapter, and we zoom in on the networked organization and the virtual organization as two alternative forms of organizing that might require a different vocabulary to describe their operations. This leads us to bring in the concept of 'infrastructure' in order to account for the technological, material, and symbolic elements that are central to organizing. In analyzing digital organizing, we argue, the concept of structure remains useful, but supplementing it with the concept of infrastructure helps us capture organizational phenomena that are important for any organization, but maybe especially those being subjected to digitalization. The chapter tells the story of the digitalization of a library. It illustrates how digital organizing alters an organization's relations with its users. Its formal structures are not changed, but less visible *digital infrastructures* pose a challenge to the formal structures, and the hierarchical relations are disturbed.

In Chapter 5, we turn to the theme of 'production' in organization studies. Here, we identify a profound shift in the way scholars and observers have perceived the organization of work and production in capitalist society. Whereas organization scholars originally maintained a sharp distinction between the activities of production and consumption, as well as between the social categories of manager and worker, more recent techno-optimist observers argue that the proliferation of digital technology has blurred these distinctions to the point where they are no longer relevant. According to them, this technology-driven fusion of roles and activities is best captured by compound words such as 'prosumption' (production and consumption) and 'produsage' (production and usage). The chapter argues that while such concepts draw our attention to important developments in the field of digital organizing, they also blind us to situations where organizational roles and activities continue to remain distinct. When we conflate categories like 'the producer', 'the consumer', 'the worker', and 'the manager', we surrender our ability to meaningfully identify those actually producing and consuming commodities in the digital economy, as well as those profiting from this process. We thus emphasize the value of retaining some of the original meaning of organized production while recognizing the new dynamics captured by concepts such as 'produsage'. As an illustration, the chapter discusses the sale of Huffington Post to America on Line (AOL). It led to public debates and lawsuits when the then editor-in-chief, Arianna Huffington, turned a vibrant community of voluntary freelance writers into big business. Her profits were huge, but the thousands of freelancers producing content for the site received nothing. The reason was that bloggers had contributed voluntarily to the portal and were not under contract – we could call them 'produsers'.

Chapter 6 highlights how the field of organization studies has been concerned with the problem of motivating individuals to collaborate toward a common goal, and how scholars have understood collaboration among people and among organizations. From this literature, we understand that face-to-face encounters, physical proximity, and digital technologies all affect the way collaboration takes place. With the arrival of digital technologies, both intra-organizational and interorganizational collaboration can be supported in new ways. Digital technologies have also generated new platforms for co-creation, crowdsourcing, and open innovation. The chapter argues that co-creation – and related concepts such as crowdsourcing – can be seen as reflecting expectations regarding openness and involvement on the part of users, customers, and citizens, expectations stimulated by advancements in digital technologies. The chapter shows how the concepts of collaboration and co-creation, while different, are not mutually exclusive in organizational practices. The central case in the chapter is the launching of a co-creation platform where magazine journalists and readers contribute to producing feature stories. It shows some of the changes in journalism, but it also shows that traditional journalistic collaboration still takes place. In that way, co-creation is not *replacing* collaboration.

In organizations – and in organization studies – both knowledge and data are considered resources that can help organizations work smarter, deliver higher quality, and/or gain competitive advantage. Knowledge is the topic of Chapter 7. Knowledge has a longer history than data as a matter of concern for organizations, and specialists in knowledge management have developed various techniques for 'capturing' and sharing knowledge.

In this tradition, people are considered bearers of implicit knowledge. Organizations' concern with knowledge management has intensified over several decades, but advances in digital technologies have provided new conditions for these practices. Originally, digital technologies functioned as repositories for saving and sharing information. Today, the production of knowledge has become intertwined with new ways of using and handling data – for instance through the use of AI or automated treatment of large data sets. Data has acquired another status – it is no longer the lowest rank in the knowledge hierarchy, below 'information' and 'knowledge'. In everyday talk, data is portrayed as a force of its own. Data now has agency, as expressed in terms such as 'data-driven management'. In the chapter, we speak of *datafication* in order to highlight this qualitatively different knowledge paradigm in organizations. We discuss how knowledge and datafication are both considered crucial resources to be harvested and managed, and how they facilitate the roles of people and technology in different ways. We illustrate these points with a story of the introduction of mobile devices in home care. The quality of the care provided is crucially dependent on how knowledge about clients is produced, stored, used, and exchanged. The new mobile devices produce large quantities of extra data, and data acquire a new role in the organization. The case shows that both more traditional knowledge management and the handling of new types of data are important to the organization.

In Chapter 8, we describe how communication in organizations has been practiced and theorized. Here we introduce the term 'interactivity' as a useful concept in our understanding of communication and as an empirical phenomenon afforded by digital technologies. The chapter highlights some important traditional subdisciplines and concepts of organizational communication such as internal and external communication, strategic communication, and corporate communication. These disciplines and concepts have been challenged by the complexity of communication in practice, and by theoretical developments highlighting the narrative dimension of communication and the material elements that also impact communication. All these developments leave us with an image of organizational communication as increasingly difficult to carry out, and to control, particularly in light of the communication practices afforded by digital technologies. The chapter proposes that along with organizations' ability to analyze and conduct communication as a planned and tightly managed activity, organizations will be increasingly required to master the interactive aspect of communication. Expectations about interactivity and new digital platforms that allow for unprecedented interactivity create new conditions for organizational communication. We illustrate this point with a story of organizational communication on Twitter, where both members and non-members of the organization play a part in constituting its reputation. The case exemplifies how social media has made the old style, one-way communication nearly impossible. Even if companies spend resources on tightly planned strategic communication, they need to react to the fact that we live in a world of 'interactivity'.

Chapter 9 describes how legitimacy has become a central concern for organizations, and how this is often linked to demands for more transparency – especially because expectations of transparency have grown with the advent of digital communication technologies. The chapter opens with a historical account of how it has become increasingly important whether an organization and its activities are viewed as legitimate from a societal point of view, and how legitimacy has been treated in organization studies. The chapter then describes how legitimacy is increasingly tied to transparency. We zoom in on the concept of transparency, first by discussing the simple understanding of transparency as a practice of delivering 'more and better information'. We then move to a discussion of how transparency is generated in organizational practices. From our discussions throughout this chapter, it becomes clear that legitimacy is not something organizations 'have' and that transparent is not something organizations 'are'. Both phenomena are dependent on the way in which people and technologies interact. The central argument of the chapter is that digital technologies affect how both legitimacy and transparency are constructed in organizations. We treat them as interconnected phenomena because digital technologies create new expectations of – and means of producing – transparency. We illustrate this point with the story of an oil company that made use of a corporate blog to work strategically with legitimacy and transparency.

In Chapter 10, we argue that power is not a uniform concept, even though the techno-optimist literature on digital organizing has treated it as such. As a corrective to this view, we discuss four dimensions of power in organizations – coercion, manipulation, domination, and subjectification. All four views of power can be traced throughout organizational studies. Many scholars now acknowledge that power is a multidimensional phenomenon in organizations. However, the techno-optimist literature often deploys a more 'intuitive' understanding of power as coercion. In this light, they end up speaking of 'empowerment' in unambiguously positive terms. When power is viewed as a repressive force that prevents people from pursuing their real interest, empowerment is seen as a force of liberation. And since digital technology is viewed as a driver of individual autonomy, digitalization comes to be equated with widespread empowerment of ordinary people. There is a weakness in the techno-optimist literature, however. It tends to overlook the fact that power can work through the 'hearts and minds' of people. As such, the chapter proposes that studies describing empowerment in digital organizing can fruitfully be supplemented with a multidimensional view of power. The chapter discusses the case of digital freelance platforms. Such platforms offer opportunities to escape the constraints of a traditional work relationship: freelancers are empowered because no one is forcing them to do anything. At the same time, such platforms also require that the freelancer engages in promoting an image of being industrious and systematic, competitive and trustworthy, self-confident and friendly. Here, power relations work 'through' the individual.

Together, Chapters 4 through 10 show how digital technologies facilitate changes in a range of organizational practices and relations, although we must remember that digitalization neither determines these practices and relations, nor totally alters them. The intent of these chapters is to show that it makes little sense to say that digital technologies provoke a move 'from power to empowerment' or 'from communication to interactivity'. Instead, these phenomena can be examined as two sides of the same organizational coin. The chapters demonstrate that it is possible to speak about digitalization in organizational terms, by engaging with concepts and concerns from a broad range of organizational themes. Each of the thematic chapters ends by posing a set of questions regarding the types of organizational analyses that could be relevant in a given domain and the kinds of issues that arise in digitalized organizations.

The final chapter takes a step back. Here we reflect upon some of the wider implications of digital organizing for individuals, organizations, and society. We do not attempt to speculate about the future implications of digital organizing, as such speculations often say more about the present than the future (Andersen & Pors, 2016). However, we will point out some of the visible effects of digital organizing at the levels of individuals, organizations, and society. We do so in order to invite reflections about the normative, ethical, and political aspects of digital organizing. The digitalization of organizations is not merely a technical project. It is precisely because digitalization is intertwined with non-technical aspects that it requires human thoughtfulness and responsibility.

? QUESTIONS

1. How does binary thinking underlie organizational restructuring?
2. How do visions related to exponentiality inform organizational choices and investments?
3. How do organizations grapple with the tension between openness and control in a digitally connected organizational reality?
4. How is automation bounded in specific organizations?
5. How is automation blended in a given organization?
6. How are classical themes in organization studies relevant to the analysis of digital organizing?
7. Why does digital organizing require new analytical concepts?
8. What are some important normative, ethical, and political aspects of digital organizing?

References

Andersen, N. Å. & Pors, J. G. (2016). *Public management in transition: The orchestration of potentiality*. Bristol: Bristol University Press.

Beunza, D. & Millo, Y. (2015). Blended automation: Integrating algorithms on the floor of the New York Stock Exchange. *SCR Discussion Paper No. 38*. London: London School of Economics. Retrieved from http://www.systemicrisk.ac.uk/sites/default/files/downloads/publications/dp-38.pdf

Boole, G. (1854). *An investigation of the laws of thought: The mathematical theories of logic and probabilities*. Cambridge: Macmillan and Co.

Brown, J. S., Weinberger, D. & Duguid, P. (2017). *The social life of information*. Cambridge, MA: Harvard Business School Publishing.

Brynjolfsson, E. & Mcafee, A. (2014). *The second machine age: Work, progress, and prosperity in a time of brilliant technologies*. New York: W.W. Norton & Company.

CERN (2019). A short history of the web. Retrieved from https://www.google.com/search?client=safari&rls=en&q=tim+berners+lee+CERN&ie=UTF-8&oe=UTF-8

Christensen, C. M. (1997). *The Innovator's dilemma: When new technologies cause great firms to fail*. Cambridge, MA: Harvard Business Review Press.

Clegg, S. (2014). Organisationsteori: Et historisk overblik. In S. Vikkelsø & P. Kjær (Eds.), *Klassisk og moderne organisationsteori* (pp. 11–22). Copenhagen: Hans Reitzels Forlag.

Du Gay, P. & Vikkelsø, S. (2017). *For formal organization: The past in the present and future of organization theory*. Oxford: Oxford University Press.

Fleming, P. (2019). Robots and organization studies: Why robots might not want to steal your job. *Organization Studies, 40*(1), 23–38.

Ford, M. (2015). *Rise of the robots: Technology and the threat of a jobless future*. New York: Basic Books.

Galloway, S. (2018). *The Four: The hidden DNA of Amazon, Apple, Facebook and Google*. New York: Penguin.

Google for Education (2017). Exploring computational thinking. Retrieved from https://edu.google.com/resources/programs/exploring-computational-thinking/

Graham, S. & Thrift, N. (2007). Out of order: Understanding repair and maintenance. *Theory, Culture & Society, 24*(3), 1–25.

Gupta, M. (2018). *Blockchain for dummies*. Hoboken, NJ: John Wiley & Sons.

Iansiti, M. & Lakhani, K. R. (2017). The truth about blockchain. *Harvard Business Review, 95*(1), 118–127.

Justesen, L. & Plesner, U. (2018). Fra skøn til algoritme: Digitaliseringsklar lovgivning og automatisering af administrativ sagsbehandling. *Tidsskrift for Arbejdsliv, 20*(3), 9–23.

Kamstrup, A. (2017). Crowdsourcing and the architectural competition as organisational technologies. Doctoral dissertation, Department of Organization, Copenhagen Business School.

Latour, B. (2013). What's the story? Organizing as a mode of existence. In D. Robichaud & F. Cooren (Eds.), *Organization and organizing: Materiality, agency, and discourse* (pp. 37–51). New York and London: Routledge.

Law, J. (2004). *After method: Mess in social science research*. London: Routledge.

Lepore, J. (2014). The disruption machine: What the gospel of innovation gets wrong. *The New Yorker*, June 23. Retrieved from https://www.newyorker.com/magazine/2014/06/23/the-disruption-machine

O'Neil, C. (2016). *Weapons of math destruction: How big data increases inequality and threatens democracy*. New York: Penguin Books.

Plant, S. (2016). *Zeros + ones: Digital women and the new technoculture*. London: Fourth Estate.

Plesner, U., Justesen, L. & Glerup, C. (2018). The transformation of work in digitized public sector organizations. *Journal of Organizational Change Management, 31*(5), 1176–1190.

Provenance (2018). Every product has a story. Retrieved from https://www.provenance.org/

Riordan, M. & Hoddeson, L. (1997). *Crystal fire: The birth of the information age*. New York: Norton.

Roose, K. (2019). The hidden automation agenda of the Davos Elite. *The New York Times*, January 25. Retrieved from www.nytimes.com/2019/01/25/technology/automation-davos-world-economic-forum.html

Scholz, T. M. & Stein, V. (2018). The architecture of blockchain organization. *ICIS 2018 Proceedings*.

Singularity University (2018). Preparing global leaders and organizations for the future. Retrieved from https://su.org

Smith, A. (2018). Franken-algorithms: The deadly consequences of unpredictable code. *The Guardian*, August 30. Retrieved from https://www.theguardian.com/technology/2018/aug/29/coding-algorithms-frankenalgos-program-danger

Susskind, R. E. & Susskind, D. (2015). *The future of the professions: How technology will transform the work of human experts*. Oxford: Oxford University Press.

Wajcman, J. (2017). Automation: Is it really different this time?. *The British Journal of Sociology, 68*(1), 119–127.

World Wide Web Foundation (2019). History of the Web. Retrieved from www.webfoundation.org.

Zuboff, S. (2019). *The age of surveillance capitalism: The fight for a human future at the new frontier of power*. New York: PublicAffairs.

2 TECHNOLOGY AS A THEME IN ORGANIZATION STUDIES

> In this chapter, we explore the overall theme of technology in organization studies. We do so by chronicling relevant developments in organization theory and then linking these to the study of different types of technology. Organization studies was initially a rather technical science closely related to the science of engineering. However, it gradually lost sight of technology as organizational scholars shifted their attention to 'softer' areas. Therefore, we need a vocabulary that will enable us to integrate technology into organizational understanding without losing the insights and perspectives that have enriched the discipline as it evolved. As we show, the notion of technology has been used in numerous ways. For some, the term denotes particular practices and rationalities of governing. Others use the term 'technology' in a more conventional sense to describe technological artifacts. The chapter's discussion of classical approaches to the question of technology in organizations offers a backdrop for analyzing digital organizing. As such, this chapter – as well as the entire book – connects studies of present-day digital technologies to the field of organization studies.

Organization studies as a technical science

While the American political scientist Luther Gulick (1937) was the first to articulate an actual 'theory of organizations', most scholars trace the birth of organization theory to the emergence of the Scientific Management movement, which later became the backbone of classical organization theory. Scientific management is often associated with Frederick Winslow Taylor's (1911) influential essay *The Principles of Scientific Management*, which describes how managers can make manufacturing more efficient and productive. One of Taylor's core arguments was that the best way to increase productivity in industrial manufacturing was to establish a rigid division of labor between managers and workers. Managers would apply scientific methods to decide how the workers on the shop floor should carry out their tasks. In fact, Taylor even suggested removing all kinds of 'brain work' from the shop floor, thereby reducing the manual worker to a simple cog in the machine of industrial organization (Taylor, 1903/1919, p. 36).

From reading Taylor's accounts of scientific management and other landmark texts from the early days of organization theory such as Harrington Emerson's (1912) *The Twelve Principles of Efficiency* or Henri Fayol's (1916) *General and Industrial Management*, one might conclude that the question of technology was basically left unaddressed. Efficient management was simply about re-organizing the labor process. However, although classical organization theorists rarely dealt with technology as such, the concept played an important role in establishing organization studies as a technical science closely related to the science of engineering (Shenhav, 1995). One example of scientific management's close link to engineering is the fundamental assumption that all parts of an organization's production system – technical as well as human – should be interchangeable. In other words, the theory suggests that just as technical components can be changed if they stopped operating according to expectations, manual workers can be replaced if their productivity suddenly declines. Scientific management scholars pursued the virtues of efficiency and productivity by trying to standardize both the technical processes and human work tasks (Korver, 1996). As Taylor, an engineer by training, states: "It would seem almost unnecessary to dwell upon the desirability of standardizing, not only all of the *tools*, appliances and implements throughout the work and office, but also the *method* to be used in the multitude of small operations which are repeated day after day" (Taylor, 1903/1919, p. 123, italics added).

Taylor and other members of the Scientific Management movement such as Frank and Lillian Gilbreth, grounded their quest for standardization in 'time-motion' studies (Kijne, 1996). The basic idea of time-and-motion studies is to measure the exact movements required to carry out a task and the time it takes to complete that sequence of movements, with the aim of identifying potential efficiency gains. For instance, in *The Principles of Scientific Management*, Taylor recounts his own efforts to increase efficiency among a group of workers handling pig iron at the Bethlehem Steel Company in Pennsylvania. Through careful measurement, Taylor observed that the workers managed to load only 12.5 tons of iron per man per day, even though this figure was supposed to be somewhere between 47 and 48 tons. To deal with this inefficiency, Taylor summoned the factory's best worker, a man called Schmidt, and convinced him to work under the close supervision of a man with a watch who would direct his every move: "Now pick up a pig and walk. Now sit down and rest. Now walk – now rest" (Taylor, 1911/1967, p. 47).[1] By applying similar methods of 'enforced standardization', Taylor managed

---

1 The dialogue between Taylor and Schmidt vividly illustrates Taylor's condescending attitude toward manual workers. For instance, to emphasize the "mentally sluggish" nature of men like Schmidt, Taylor deliberately incorporated spelling errors into Schmidt's part of the dialogue (Taylor, 1911/1967, pp. 44–46).

to increase each worker's efficiency. By the end of Taylor's tenure, all pig-iron handlers were able to load almost four times the amount of iron. The apparent success of such management techniques led Taylor to ridicule the many "good managers of the old school" who perceived standardization as both unnecessary and insidious (Taylor, 1903/1919, p. 124). According to Taylor, these managers seemed to prefer that workmen should develop their individual personalities by choosing the work methods they felt best suited them. However, allowing manual workers such liberties made little sense to Taylor and his colleagues in the Scientific Management movement, as such freedom would presumably reduce efficiency.

The theory of scientific management received an extraordinary amount of attention from scholars and industrial management practitioners around the world, and several factories adopted the methods prescribed by Taylor. In fact, the movement became so popular during the industrialization phase of the late nineteenth and early twentieth centuries (Henry Ford being perhaps the most well-known admirer) that some scholars now view it as one of the most influential theories in modern social science (Merkle, 1980). However, at the peak of its fame in the early twentieth century, harsh critiques also began to emerge, most of which focused on the resulting degrading working conditions that 'Taylorism' generated for manual workers. After a strike at the Watertown Arsenal factory in Boston and a subsequent hearing in the US House of Representatives, the popularity of scientific management began to fade. Taylorism was now associated with the exploitation of the worker and his alienation from the product of his labor. The intense focus on standardization and specialization, critics argued, dehumanized the manual labor force by depriving its members of the right to perceive themselves as controllers of their own labor process (Kjær, 2014).

Charlie Chaplin famously mirrored the critique of scientific management's dehumanizing propensities in the 1936 film *Modern Times*, in which the hard-working protagonist spends every workday tightening bolts on an assembly line. As a result of the repetitive nature of this tedious task, his movements become increasingly mechanical. Eventually, after jumping onto the assembly line in order to tighten the bolts faster, the exhausted worker becomes completely entangled in the factory's machinery – it becomes a very literal visualization of the worker as a cog in the machine of industrial manufacturing (Image 2.1).

Although Taylorism has been heavily criticized, it is frequently invoked in present-day accounts of the digitalization of organizations. This is because digital technologies render many work processes visible and measurable in new ways. On this basis, even administrative work processes can be divided into individualized segments and managed in a compartmentalized way. When an organization bases its workflows on shared access to data and digital infrastructures, employees do not necessarily have 'their own' customers or administer 'their own' cases. Instead, it is argued, they become replaceable components in the digitalized organization.

TECHNOLOGY AS A THEME IN ORGANIZATION STUDIES

Image 2.1 The interior of a gold stamp mill in 1888. Beginning in the late eighteenth century, factories emerged as a new type of workplace. The assembling of large numbers of workers around production technologies gave rise to early organization and management theories such as Taylor's scientific management.

Source: Library of Congress/Science Photo Library, Ritzau/Scanpix.

The human factor

Until the 1930s, classical organization theory was primarily dominated by two lines of thought: scientific management, as described above, and the bureaucratic organization approach most famously outlined by the German sociologist Max Weber (1922).[2] In contrast to scientific management's pursuit of profit and efficiency, Weber saw bureaucracy as a rational and ethical way of organizing. Despite various differences between them, the two perspectives shared an interest in formal structures and clear lines of command. It was not until the arrival of the so-called Human Relations School that there emerged a challenge to classical organization theory. Scholars associated with the Human Relations School, instead of focusing strictly on formal organization, developed an interest in organic structures and sociopsychological processes. Consequently, organization studies

---

2 Most of Weber's work was not translated into English until the late 1940s, but his writings had a great influence on German-speaking readers.

became less preoccupied with technical concerns and more interested in what became known as the 'human factor' (e.g. Schein, 1996).

The Human Relations School is often associated with a famous study conducted over the course of five years (1927–1932) by a team of researchers at the Hawthorne Works, an Illinois-based factory owned by the Western Electric Company (Roethlisberger & Dickson, 1939). Along the same lines as the scientific management tradition, the Hawthorne experiments began as an inquiry into tangible ways of increasing productivity in industrial manufacturing. From the outset, the guiding hypothesis was that a causal relationship existed between workers' productivity and the quality of the physical environment such as the lighting at the factory. The researchers hypothesized that worker productivity would increase if the lighting quality was increased. They therefore began a series of 'illumination experiments' in which they increased the wattage of the light bulbs in three separate departments at the factory. As expected, the productivity level rose, and the hypothesis appeared to be confirmed. However, the productivity gains were not as linear as the researchers expected, so the researchers decided to conduct two more experiments.

The results of the first additional experiment were equally ambiguous, as the researchers could not determine the exact relationship between lighting and productivity. They therefore began to speculate that the amount of daylight was distorting their experiments. This led them to invite a group of female workers to work in a separate assembly room without daylight. The workers were asked to continue their job of assembling relays, while the room's artificial lighting was increased and decreased according to plan. At this point, the researchers began to speculate that psychological processes, rather than technical factors, were causing the spikes in productivity that had been detected in the previous experiments. Therefore, they decided to trick the workers: they decreased the amount of light while telling them that they were, in fact, increasing the amount of light. Notably, the level of productivity continued to rise. These discoveries led the researchers to conclude that the productivity gains had been caused not by any increase in light wattage but by the increased amount of attention paid to the workers by the factory managers and the research team. The group of workers no longer saw themselves as ordinary workers in a large American factory. Instead, they had begun to see themselves as an elite unit referring directly to both researchers and management, and it was this new relationship that apparently caused their productivity to increase. Today, this phenomenon is widely known as the 'Hawthorne effect' (Gillespie, 1991).

After presenting Western Electric Company with their conclusions, the researchers decided to delve deeper into additional factors affecting workers' productivity at the Hawthorne Works. Inspired by their lighting experiments, they launched a much larger study of human relations in industrial organization. Studying everyday practices at the factory, they began by interviewing more than 21,000 workers and closely monitoring a group of male workers whose job consisted of mounting wires on telephone equipment. From the vast amount of data they collected, the researchers concluded that informal relationships among workers and unwritten rules within work groups had much greater effects on workers' productivity and motivation than previously assumed. Prior to the study, both factory management and the academic community had assumed that wages and, possibly, lighting levels were the defining factors influencing productivity. However, the researchers demonstrated that it was the 'social organization' of work groups which exercised decisive control over group members' behavior and, hence, the overall productivity on the shop floor (Roethlisberger and Dickson, 1939, p. 380).

The Hawthorne studies were important for the theme of technology within organization studies. But this was not because of their results, which indicated that technical factors (i.e. lighting) were less important than human factors. Rather, it was because of the universalism that was later attached to their conclusions. As several scholars have attempted to show, the conclusions advanced by the Hawthorne researchers, most notably Elton Mayo, were presumably driven by an ideological desire to advance sociopsychological studies of organizations and to discredit the technical nature of classical organization theory (Gillespie, 1991; Hassard, 2012). In Mayo's (1933) account, for instance, technical factors were assigned remarkably little importance, while human factors such as social norms and group belonging were at the heart of his conclusions. The researchers' clear emphasis on the importance of human relations helped move organization studies away from its heritage as a technical science and, therefore, away from the issue of technology (see Figure 2.1).

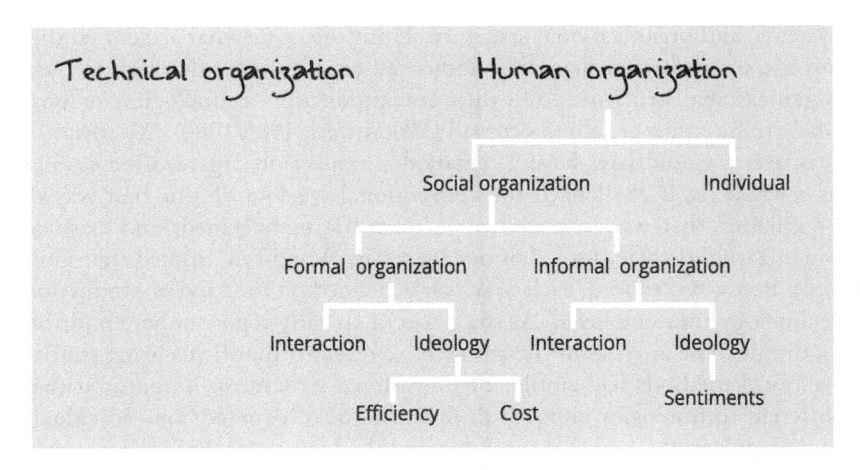

Figure 2.1 The industrial plant as a social system. Visualization of Fritz Roethlisberger and William J. Dickson's (1939, p. 565) summary of "the different parts into which an industrial plant as a social system can be divided". Notice how elaborately the 'social organization' is described compared to the 'technical organization' and the 'individual'.

Technology takes center stage

Decades later, the theme of technology was almost accidentally reintroduced into organization studies by the British industrial sociologist Joan Woodward. As Charles Perrow (1973, p. 9), a well-known organization theorist, later explained, Woodward "tripped and stumbled over a piece of gold ... She picked up the gold and labeled it 'technology', and made sense of her otherwise hopeless data." Woodward discovered the gold in 1953 when she decided to subject the management theories that she was teaching at a college in south-east Essex to empirical testing (Dawson & Wedderburn, 1980). To this end, Woodward summoned a team of researchers and began to map out the field of nearby industrial plants. The team eventually settled on 100 firms with more than 100 employees and commenced their research. Like so many scholars before her, Woodward and her colleagues initially tried to identify 'one best way of organizing' by searching for some kind of single pattern across the firms. Her team collected massive amounts of data about each firm's history, manufacturing processes, structures, work routines, and financial results (Woodward, 1965/1980, p. 11). As the investigation progressed, however, Woodward realized that she was unable to detect any generalizable laws of efficient organization. It was simply impossible to empirically determine exactly which structural arrangement (in the singular) provided the best foundation for commercial success in industrial organization.

Instead of abandoning the study, Woodward and her colleagues took one last look at the data. They examined one variable that had not previously received much attention – the firms' core production technologies. Suddenly, a pattern emerged. Through a close re-reading of the data, Woodward discovered a causal relationship between what she called 'production systems' and organizational structure. From this somewhat accidental discovery, she theorized that those industrial organizations that design their organizational structures to fit their core production technologies are most likely to be commercially successful (Woodward, 1965/1980). Woodward's 'discovery' would later have a remarkable impact on organization studies as a whole, as it challenged the conventional wisdom of 'one best way of organizing' that was so prevalent in scientific management and classical organization theory. Instead of one best way, Woodward argued that multiple 'best ways' existed, each of which depended on the kind of production technology they employed. As she states in an early report of her empirical findings: "The analysis of the research ... revealed that firms using similar technical methods had similar organizational structures. It appeared that different technologies imposed different kinds of demands on individuals and organizations, and that these demands had to be met through an appropriate form of organization" (Woodward, 1958, p. 16).

Woodward's main idea was to divide the 100 firms into three categories according to their core production technology: (1) *unit and small-batch firms* (e.g. firms that produced tailor-made clothing or customized furniture), (2) *large-batch and mass-production firms* (e.g. car manufacturers and large bakeries), and (3) *process production firms* (e.g. pharmaceutical companies with research and development units). After isolating technology as a variable, Woodward observed that firms with similar production technologies had similar organizational structures. For instance, process production firms were generally much more decentralized than firms active in mass production, and small-batch firms tended to have far fewer management levels than process production firms. Woodward also found that those firms that clustered around the median within each category tended to be most successful commercially (Woodward, 1965/1980, pp. 50–52). Furthermore, through an analysis of 23 case studies, Woodward's research team was able to develop a series of prescriptive propositions about the most common manufacturing cycle for each category. In small-batch firms, for instance, the manufacturing cycle went from marketing to development to production. In large-batch firms, the cycle began with development before proceeding to production and finally marketing. The most critical function, Woodward asserted, was the one in the middle. Hence, development was the most critical function for small-batch firms, production was the most critical function for large-batch firms, and marketing the most critical function for process production firms.

While some of the arguments contained in Woodward's seminal work, *Industrial Organisation*, might seem somewhat trivial to today's reader, they were strikingly provocative at the time of publication. The provocation had to do with Woodward's critical stance toward the canonical texts within organization studies. Perhaps for that reason, her writings were extensively criticized. While some of the critics claimed that her research lacked sound methodological considerations (she rarely provided details about her analytical strategy or how she reached certain conclusions), others noted that her principal argument bordered on technological determinism. Despite the critiques, Woodward's claims of a causal relationship between technology and organizational structure have had a profound influence on organization studies in the latter part of the twentieth century. In fact, the school of thought that became known as 'contingency theory' is often said to have grown out of Woodward's techno-sensitive perspective (Donaldson, 2001). However, before we address contingency theory, we need to consider an often neglected but very important theory of organization: 'sociotechnical systems theory', or simply 'sociotechnical theory'.

Sociotechnical theory

In the early 1950s, when Woodward was beginning her fieldwork, a group of British scholars associated with the Tavistock Institute of Human Relations began to investigate the relationship between social and technical elements within organizations. Their main area of interest was the potentially disruptive effects of new technologies, especially when they were introduced into well-established work organizations. Notably, for these researchers, the term 'technology' does not necessarily denote something of a purely material nature. For instance, one of the most well-known studies to come out of sociotechnical systems theory explored the introduction of a new production system called the Longwall Method at a coal mine in Durham, England (Trist & Bamforth, 1951). The Longwall Method is a coal-retrieval system that allows colliers to extract long walls of coal instead of short faces. In many respects, this was a significant improvement in terms of efficiency and workers' safety. However, the introduction of the Longwall Method entailed a degree of mechanization of otherwise purely human work. Tasks that had been carried out by hard-working colliers were now suddenly carried out by machines, and this transfer of labor power turned out to have crucial consequences for the psychological well-being of the colliers.

Prior to the introduction of the new technology, teams of two colliers (the hewer and his partner) would carry out all the tasks required to extract coal from the mine such as hewing the coal, loading it into tubs, and ripping the ceiling. In principle, all workers knew how to conduct all tasks. This was called the Hand-Got System. As mining often occurred deep below the surface, the Hand-Got System operated primarily on the basis of the miners' own self-management. The system gave the colliers a high degree of autonomy and placed all the responsibility on the shoulders of the individual groups. This type of work organization, structured around the notion of 'responsible autonomy', freed management from the tedious and resource-heavy task of supervising subordinates, while it simultaneously provided the individual colliers with a certain "craft pride and artisan independence" (Trist & Bamforth, 1951, p. 6). In addition to the pride of working in semi-autonomous groups, the Hand-Got System created a strong bond between team members. Team members cared for each other, both professionally and personally. When the Longwall Method was introduced, much of this pride and social cohesion declined.

Through two years of extensive fieldwork in and around the Durham mine, Eric Trist and Ken Bamforth investigated how the Longwall Method not only mechanized a series of tasks but also deprived the colliers of their responsible autonomy. In contrast to the previous situation in which each collier served as a generalist who could carry out all the tasks involved in mining coal, the Longwall Method separated the workers into well-defined occupational roles with corresponding wages. Furthermore, the many small groups of the Hand-Got System were merged into larger groups of 40 workers, with one group doing the cutting, another group doing the ripping, and a third group handling the filling. Therefore, this new technology broke down the close relationship between the hewer and his partner, replacing it with more efficient production, but also more impersonal relationships. On the psychological level, the introduction of the Longwall Method generally caused the workers to become more anxious, stressed, and isolated. As they were now at the mercy of machines without much opportunity to control the labor process, the workers experienced feelings of apathy and alienation, much akin to the experience of assembly line workers subjected to the principles of scientific management. The colliers responded by forming informal groups that often engaged in counter-productive activities such as 'mutual scapegoating' and 'self-compensatory absenteeism' (Trist & Bamforth, 1951, p. 33).

Trist and Bamforth's work at the Durham coal mine spawned a series of studies focused on the technical and sociopsychological repercussions of introducing new technologies into different kinds of work organizations. For instance, A. K. Rice (1953) investigated the introduction of half-autonomous work groups at an Indian textile mill. Later, Fred Emery and Einer Thorsrud (1969) examined novel processes of organizational democratization in the Norwegian manufacturing industry. These researchers highlighted the fact that technologies such as the Longwall Method should not be viewed as purely technical systems consisting of cogs and wheels, or nuts and bolts (as Woodward might have argued). Nor are they purely social systems derived from informal social relations (as Mayo might have proposed). Instead, these researchers suggested that 'technology' should be seen as itself a product of two systems converging – as a sociotechnical system (Kelly, 1978). Accordingly, the Tavistock researchers argued that neither technical nor human factors determine the character of organizations. Hence, the 'best way' of organizing was that of balancing technical and social factors in a provisional state of equilibrium. As such, sociotechnical systems theory can be seen as a way of merging Woodward's techno-sensitive perspective with the more humanistic insights of the Human Relations School, spearheaded by Mayo and the Hawthorne researchers. We should keep in mind the insights of sociotechnical systems theory when, later on, we discuss the socio*material* perspective, which is often invoked in analyses of digital organizing. Both the sociotechnical and sociomaterial approaches have the same project: trying to integrate the social and technological/material elements of organizing. The sociomaterial perspective is, however, a bit more radical in asserting that the two dimensions are constitutively entangled. We will get back to a discussion of the sociomaterial approach in Chapter 3.

Moving toward contingency

In the early 1960s, the theme of technology was once again revived by a new wave of organization theorists. While they were initially inspired by the writings of Woodward and members of the Tavistock Institute, this new wave of scholars began to question the central position that Woodward and Tavistock had assigned to technology. While Woodward's main argument could be viewed as close to some form of technological determinism (see Chapter 3), the Tavistock researchers maintained that technical factors were only one of two separate systems that determined the structure and dynamics of organizations (to be fair, some sociotechnical theorists viewed the economy as a third system). Nonetheless, the new group of scholars felt that this relatively neat way of conceiving organizations was too simplistic. The reality of organizing, they maintained, could not be described by isolating only one or two variables. Organizations were messier than this perspective suggested. For scholars like Paul Lawrence, Jay Lorsch, and Charles Perrow, this messiness was best encapsulated by the notion of contingency.

According to the Oxford English Dictionary, the word 'contingency' has two meanings: a close relationship between two or more parts, or the uncertainty of predicting future events. Although these two meanings might seem somewhat opposed, they represent two central facets of what became known as the contingency theory of organizations, or, perhaps more tellingly, the situational approach. The most basic premise of contingency theory is that "organizational effectiveness results from fitting characteristics of the organization such as its structure, to contingencies that reflect the situation of the organization" (Donaldson, 2001, p. 1). In other words, contingency theorists claim that a multitude of factors in the organization's environment determine its structure and dynamics. These environmental factors are usually referred to as 'situational variables' (Knudsen & Vikkelsø, 2014). For instance, one situational variable might be the task(s) that an organization is supposed to solve, while another could be the technology it uses. Successful organizations are able to match their internal characteristics with the situational variables. Mary Jo Hatch and Ann Cunliffe compare contingency theory to a jigsaw puzzle: "Each piece has to be shaped to fit the others as well as displaying the appropriate part of the overall picture. Unless this happens, the puzzle will be impossible to solve. Contingency theorists try to identify the key contingencies in each situation (the pieces of the puzzle), and try to determine the best fit between them" (Hatch & Cunliffe, 2006, p. 41).

The most fundamental premise of contingency theory, known as the 'fitness hypothesis', is that a change in a situational variable (x) will have an effect on the structural variable (y). The point is that organizations, like species in Charles Darwin's theory of natural selection, must constantly adapt to the changing environment in order to survive. Survival is not a matter of being big or small, rich or poor. It is a matter of being fit, because only the fittest will survive. Contingency theory thus assumes a causal relationship by which situational variables determine structural variables, not the other way around (see Mintzberg, 1980, for an exception to this one-way causality within contingency theory). The task for students of organizations, therefore, is to figure out how organizations can best adapt to certain circumstances (Figure 2.2). Stretching it a little, it can be argued that the contingency perspective overlaps with the contemporary focus on agile organizing and sprints – very common in the literature on digital technologies in organizations. Here, it is seen as increasingly problematic to build large, custom-made IT systems, which end up being outdated by the

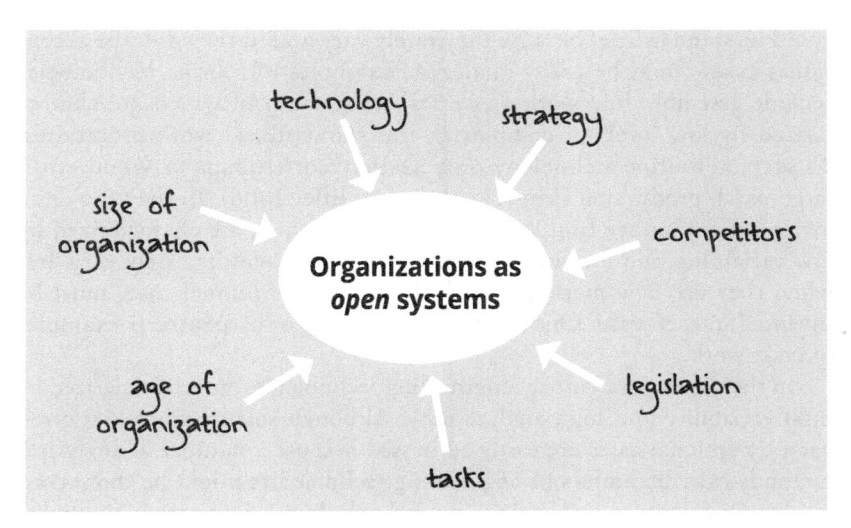

Figure 2.2 The organization as an open system, as conceived by contingency theorists. Notice how technology is depicted as one of many 'contingency factors' impacting the structure of the organization.

time they are implemented. Instead, small-scale projects can be tested continuously and adapted to technological changes and other changes in the environment.

One of the most cited studies to emerge from contingency theory is Charles Perrow's (1967) classic study of task-related technology. In contrast to Woodward, Perrow abandoned the assumption that organizations have only a single core technology. Instead, he adopted a view of organizations as determined by the tasks that they are required to carry out and the technologies they employ when doing so. As a contingency theorist, Perrow regarded technology as an independent variable and organizational structure as a dependent variable. However, his conceptualization of technology was somewhat different from most other contingency theorists. Like the Tavistock researchers, Perrow argued that technologies do not necessarily involve materiality. In fact, he defined technology as "the actions that an individual performs upon an object, with or without the aid of tools or mechanical devices" (Perrow, 1967, p. 195). After developing this definition, Perrow constructed a typology of organizational technologies by creating a two-by-two matrix (see Figure 2.3). One axis of the matrix concerns 'task variability' (i.e. the number of exceptional cases encountered when applying a certain technology), while the other concerns 'task analyzability' (i.e. the nature of the method used to deal with the exceptional cases).

This typology generated four kinds of organizational technologies. Technologies characterized by low variability and high analyzability were labeled 'routine technologies'. These were technologies (or tasks) that could be standardized because they rarely vary and, if they did, the exceptional cases could be easily analyzed. Examples of routine technologies include assembly line work or clerical work, both of which are characterized by low levels of complexity and standardized work procedures. As such, a 'routine technology organization' corresponds to Woodward's large-batch production firm (Hatch & Cunliffe, 2006). In the typology's lower left corner, we find 'craft technologies', which are characterized by low variability and low analyzability. These technologies rarely vary, but when they do, new methods for dealing with exceptional cases must be invented in each case. One might view printing or carpentry as examples of craft work.

In the upper right corner, 'engineering technologies' are characterized by high variability and high analyzability. Although such technologies often vary, exceptional cases are easily addressed because a number of analytical methods exist. Examples of engineering technologies might be those used by chemists, doctors, and engineers. Lastly, the lower right corner highlights the most complex technologies, which Perrow calls 'non-routine technologies'. Non-routine technologies are characterized by high variability and low analyzability. In these cases, exceptions occur frequently and, when they do, there are few standardized procedures for dealing with them. True to the cliché, Perrow invokes rocket science as a good example of a non-routine technology.

While few people today identify with the contingency theory label for reasons that we consider below, the theory has had an enormous impact on the way students of organizations conceive of technologies and on the definition of 'effective organization'. For instance, Lawrence and Lorsch's (1967) hugely influential study, *Organization and Environment*, is today cited more than 13,000 times, several times more than, for instance,

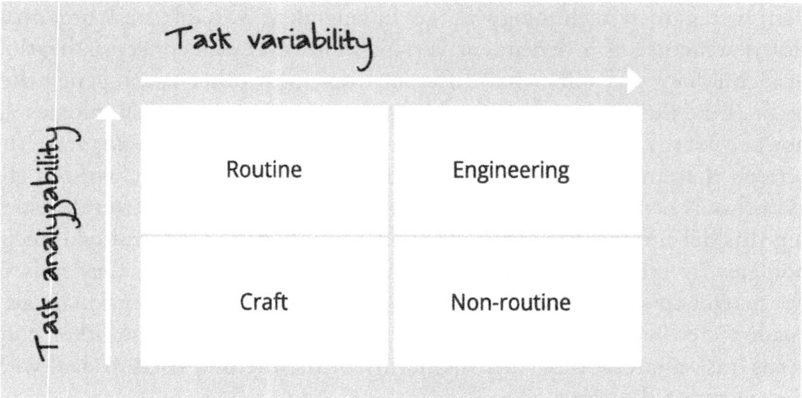

Figure 2.3 Organizational technologies categorized according to task analyzability and task variability.

The typology is modeled on Perrow (1967, p. 196).

Taylor's writings on scientific management or Mayo's account of the Hawthorne studies. In fact, if we were to name the most influential school of thought within organization studies as a whole, contingency theory would be a top candidate. Indeed, concepts of variability and analyzability are as relevant for digital organizing as they were for industrial organization, which is evidenced by discussions of how different types of work, from simple production to complex decision-making, can be replaced by digital technologies. Essentially, these are discourses of task variability and task analyzability.

Throughout the 1960s and most of the 1970s, organization theorists appeared to have finally resolved 'the technology question' in organization studies. There emerged a general consensus that although technical elements played a crucial role in determining the most suitable administrative structure, technology was only one of many situational variables that influenced the organization. For nearly two decades, this view of technology as a situational variable remained virtually uncontested. By the end of the 1970s, however, this consensus began to be questioned. The postmodern onslaught began.

Postmodernism and equivocality

Most of the studies described thus far can be seen as examples of a 'modern' style of thinking. They are primarily concerned with *end-states* rather than *processes* and with *being* rather than *becoming* (Chia, 1995). For instance, contingency theory sought to provide managers with a toolbox for understanding how to adapt to certain situations. It was less concerned with providing them with a vocabulary for understanding the process through which these situations initially emerged. However, throughout the 1970s and 1980s, the modern style of thinking was seriously challenged by what became known as *postmodernism* (see Parker, 1992). Often associated with an influential book by the French philosopher Jean-François Lyotard (1979), postmodernism ushered in a break with the modern style of thinking and the prioritization of end-states over processes. While modernist theories were structured around grand narratives derived from the Enlightenment such as rationality and reason, postmodernism promoted an "incredulity towards metanarratives" (Lyotard, 1984, p. xxiv). Rather than constructing grand theories and generalized truths, the task for postmodern thinkers was to de-construct (i.e. expose to critical scrutiny) those 'truths' that served to sustain the modern condition. Instead of enhancing our understanding of what it means to act rationally, postmodernism was (and is) concerned with criticizing and de-constructing the very notion of rationality itself (see Derrida, 1976).

Postmodernism had a significant impact on organization studies, perhaps most visibly illustrated in the work of scholars like Gareth Morgan (1986) and Stewart Clegg (1989). The hard-won consensus that contingency theory had achieved through its persistent focus on situational variables and structural fitness started to collapse, as organization theorists began questioning all they had previously taken for granted. Even the most basic concepts were scrutinized and subjected to the method of deconstruction. As Kenneth Gergen (1992, p. 703) puts it: "Why do we find it so congenial to speak of organizations as structures but not as clouds, systems but not as songs, weak or strong but not tender or passionate?" Suddenly, everything and nothing could be seen as an organization, a view that had profound consequences for the study of technology.

One of the most prominent proponents of the postmodern stance within organizations studies is the psychologist, Karl E. Weick, who published some of his most groundbreaking work a decade before Lyotard launched his attack on metanarratives. While others have placed Weick within the 'symbolic-interpretive' school (Hatch & Cunliffe, 2006), we treat him as a postmodern scholar. Weick introduced a range of well-known concepts such as 'sensemaking' and 'mindfulness' into organization studies. However, he is best known for his 1969 book *The Social Psychology of Organizing*, where he argues that there is, in fact, no such thing as organizations. At least, if organizations are conceived as discrete entities consisting of more or less static structures, they are nowhere to be found. Instead, Weick suggests, we should perceive organizations as ever-changing processes, which impels us to focus on organizing as a *verb* rather than organization as a *noun*. By attending to organizing, we can better grasp the complexity of what we commonly refer to as organizations. In the words of Weick (1969, p. 88):

> The word, organization, is a noun and it is also a myth. If one looks for an organization, one will not find it. What we will find is that there are events, linked together, that will transpire within concrete walls, and these sequences, their pathways, their timing, are the forms we erroneously make into substances when we talk about organizations.

In an effort to extend this 'process view' of organizations into different empirical fields, Weick (1990) subsequently developed a theory of technology, or, more specifically, of *new* technology. According to Weick, new technologies such as computer-driven production systems should be conceived as 'equivoque'. The notion of equivocality is central to the postmodern era of organization studies because it implies that things (i.e. technologies) do not have essential characteristics; hence, they lend themselves to potentially infinite interpretations. Etymologically, the word 'equivoque' is derived from the Latin *aequivcus*, meaning ambivalence, which is one of the defining features of the postmodern style of thinking (Kreiner, 1992). Weick refers to new technologies as equivoque because they tend to make little sense in and of themselves due to their inherent complexity and because there is so little of organizational practice related to them that is immediately visible (Weick, 1990, p. 2). One might, for instance, compare the anatomy of an old technology like the assembly line with that of a modern nuclear power plant. The inner workings of an assembly line are both visible and easy for the trained user to comprehend, and there are a limited number of exceptional cases that one might encounter while operating such a technology. A nuclear power plant, on the other hand, is much more equivoque, as the level of complexity is much higher and the possibility of understanding its complete infrastructure is limited.

In order to properly account for the equivocal nature of new technologies, Weick highlights three central characteristics that exist in older technologies but are more prominent in new technologies. Three such characteristics, which he calls 'events', stand out. They are stochastic, continuous, and abstract events. Speaking of new technologies as 'stochastic events' is a way of describing their complexity. Weick argues that new technologies tend to be insufficiently understood by those using them. This makes it hard for users to predict how and why future problems will occur, or how to solve them. This unpredictability is related not only to the complexity of new technologies but also to their constantly changing nature. Furthermore, the stochastic aspect of a new technology means that users often disagree about what it means for a technology to operate according to expectations. This uncertainty adds to the difficulty of learning from the past in order to predict future problems. As Weick notes: "Stochastic environments represent a moving target for learning because they can change faster than people can accumulate knowledge about them. When recurrence is scarce, so is learning, which is why stochastic events have become a permanent fixture of new technologies" (Weick, 1990, p. 10).

The second characteristic of new technologies Weick calls 'continuous events'. This has to do with the problem of constantly ensuring the reliability of new technologies. There is evidently some overlap between the stochastic dimension and the continuous dimension. However, while the former is concerned with the inherent unpredictability of new technologies, the latter highlights the increased need for stability and endurance. Consider, for example, a present-day stock exchange. Advanced computer technologies allow stock exchanges to operate in real time and around the clock, which is what makes stock trading a continuous event. However, this continuity creates an unprecedented need for reliability. When something goes wrong somewhere in the system, there are immediate consequences for other parts of the system. This was the case, for instance, when Knights Capital Group lost $440 million in 2012 because its automated trading system mistakenly signaled Knights Capital to purchase a significant amount of otherwise thinly traded stocks. Knights Capital Group described this incident as a 'technology issue', and it took the company a relatively long time to find out what went wrong.

Weick's final characteristic of new technologies is that they are 'abstract events'. The abstract dimension of new technology relates to visibility. For example, while one can easily observe how an assembly line works and why it sometimes stops working, doing so is far more difficult with computerized trading systems. As occurred in the case of Knights Capital Group, identifying a problem may take weeks or even months because the 'moving parts' or operative components of such systems are concealed. This demands a high level of imagination on the part of the operator/technician, which once again leaves the technology open to multiple (and often competing) interpretations (Table 2.1).

Weick's conceptualization of new technologies as stochastic, continuous, and abstract not only provides us with a vocabulary for grasping the qualities of, for instance, digital technology. It also illustrates a particular way of thinking about technology in organizations. As Weick notes, the properties of *new* technologies do not apply to new technologies alone; they can be applied to any technology. All technologies are open to multiple

Table 2.1 Equivoque technologies: Unpredictable, unreliable, and easily misunderstood?

	Characteristics	Key issue	Example
Stochastic events	New technology is so complex that it often seems to follow a random logic or pattern of development, making it difficult to predict future occurrences	Unpredictability	Driverless cars making unpredictable decisions in new situations (e.g. fatal accident in Arizona involving self-driving Uber vehicle – see Chapter 1)
Continuous events	New technology is often continuous in the sense that it operates around the clock. This creates an unprecedented need for reliability, causing operators to focus more on the maintenance of the technology than its end-product	Unreliability	Automated trading algorithms operating day and night, making close oversight necessary (e.g. Knights Capital Group losing $440 million due to 'technology issue'
Abstract events	New technology tends to be abstract because its 'moving parts' are concealed. This makes it difficult to understand what is actually going on, which is a key source of human mistakes	Misunderstanding	Value-added modeling system, assessing quality based on quantities (e.g. the firing of a capable school teacher in the US based on failure to understand an algorithm– see Chapter 1)

interpretations depending on the specific processes in which they appear. Hence, instead of exploring how technologies affect organizations in a general sense (as did Woodward or the Tavistock researchers), or how organizations can best adapt to technological changes in their environment (like the contingency theorists), the postmodern style of thinking stresses the impossibility of constructing ostensive definitions of any given phenomenon, especially technology. The postmodern scholars of organization are not trying to show that technology has no effects on organizations or humans in general. Rather, they seek to show that these effects cannot be generalized across contexts (Grint & Woolgar, 1997). Simply put, the postmodernists argue that technology has no essential characteristics. Technology is a construction, a social construction, of the relations inside organizations and between organizations and their environments.

The deconstruction of technology and other classical concepts in organization studies such as notions of 'task' and 'structure' (Vikkelsø, 2015), may partly explain why the theme of technology basically disappeared from the writings of organization theorists in the late 1980s (see Barley, 1986 and Zuboff, 1988, for important exceptions). As Raymond Zammuto and colleagues (2007) note, only 1.2 percent of the articles published in three leading journals of organization and management studies from 1996 through 2005 incorporated technology as a major theme. Having arrived at a similar conclusion, Wanda Orlikowski and Susan Scott (2008) note the paradoxical nature of this finding, given that organization studies used to be closely tied to the theme of technology. The authors suggest that one reason for the lack of technology-related research may be that organization scholars are now trained in the postmodern style of thinking that emphasizes what Robert Chia (1995) calls an 'ontology of becoming' rather than an 'ontology of being'. While the former is concerned with ephemeral processes and local micro-interactions, the latter is concerned with discrete entities and generalized truths. In the postmodern attack on metanarratives, the question of technology may therefore have been lost. Few things seem as solid and stable as technology, and what could be more modern than 'the machine'? As Keith Grint and Steve Woolgar observe (1997, p. 2): "the Enlightenment was, and modernism is, very much movements bound tightly to the idea of freedom through reason, a freedom often manifest as technology."

Technology's revival

Organization scholars working in the 1980s, 1990s, and early 2000s were to a large extent preoccupied with themes such as culture, knowledge, identity, communication, and power. Although not everyone adopted a postmodernist stance, many organization theorists abandoned the theme of technology in favor of 'softer' areas of research. Hence, exactly at a time when digital technologies associated with the internet emerged and proliferated, the field of organization studies seemed to lose interest in the question of technology. Some numbers illustrate this. During the course of only ten years, from January 1996 to January 2006, the number of internet hosting services grew by more than 4,100 percent (Zammuto et al., 2007). Similarly, the total number of mobile phone subscriptions grew from approximately 34 million in 1993 to well over four billion in 2008 (ITU, 2019). Organization scholars, however, continued to spend most of their time researching and writing about issues with no or little connection to these massive technological changes.

Nevertheless, within organizational studies and social science more broadly, a small group of scholars maintained an interest in materiality and technology. These scholars are usually associated with what became known as SST (social shaping of technology) and ANT (actor-network theory). This group of scholars shared the postmodernists' distaste for grand narratives, but their approach to technology was different. They began to explore technology as an empirical phenomenon emerging in sociomaterial networks. In so doing, they shared some interest with the techno-sensitive perspective of modernist scholars such as Woodward and the Tavistock researchers, while preserving a postmodern focus on process. The SST/ANT approach to technology was adopted by organization scholars who identified with the label sociomateriality. We will describe this approach to organization studies in more detail in Chapter 3. At this point, it is sufficient to say that sociomaterialists seek to rethink the relationship between technology and organization by rethinking the more general relationship between humans and their non-human surroundings, particularly the assumed dominance of the human over the natural or the technological. Hence, instead of privileging the social as a particularly interesting subject of analysis or as a powerful mode of explanation, these scholars focus on the inherently material nature of human existence. The view of technology and organization underpinning this book builds on such a sociomaterial approach, with a particular interest in digital technologies marked by a logic of binarity, exponentiality, and connectivity, and their interplay with organizations.

With this, our brief exposé of technology as a theme in organization studies has reached present day. We have gone into some depth with the classical studies because they are still referred to in present-day discussions of digital organizing. From this historical account, we can see that some of our concerns about technology and organization have been raised decades ago in previous theorizing (Figure 2.4). In the next chapter, we offer another set of lenses through which we can understand technology and organization. We describe three fundamentally different perspectives on how technology and organization are related, and how they influence one another.

TECHNOLOGY AS A THEME IN ORGANIZATION STUDIES

2008
'Sociomateriality' gains traction with scholars combining a techno-sensitive perspective with a postmodern style of thinking.

1979
Jean-François Lyotard launches an attack on metanarratives in his book *The Postmodern Condition*, which has significant effects on organization studies.

1967
In *Organization and Environment*, Lawrence and Lorsch are the first to describe their approach as 'contingency organization theory.'

1947
The Tavistock Institute of Human Relations is established, which marks the beginning of sociotechnical systems theory.

1927
The Hawthorne Studies are launched, which marks the emergence of the human relations school.

1996
Technology-related research in the field of organization studies reaches an all-time low, with less than 2% of the articles in one leading journal addressing the topic.

1969
Karl E. Weick publishes *The Social Psychology of Organizing* in which he argues that organization (as a noun) is a myth.

1958
Joan Woodward discovers a causal relationship between firms' core production technologies and their organizational structures.

1933
Elton Mayo publishes *The Human Problems of an Industrial Civilization*, which is later interpreted as an ideological attempt to enhance socio-psychological studies of organizations.

1911
Frederick Taylor publishes *The Principles of Scientific Management*.

Figure 2.4 Notable contributions to technology as a theme in organization studies.

? QUESTIONS

1. How is Taylor's focus on standardization still visible in contemporary organizations?
2. How might the Hawthorne studies inform present-day examples of digital organizing?
3. What are the strengths and weaknesses of contingency theory in relation to digital organizing?
4. Where does digital technology fit in Perrow's typology of organizational technology?
5. How can we study *processes* of digital organizing without losing sight of technology?

References

Barley, S. R. (1986). Technology as an occasion for structuring: Evidence from observations of CT scanners and the social order of radiology departments. *Administrative Science Quarterly, 31*(1), 78–108.

Chia, R. (1995). From modern to postmodern organizational analysis. *Organization Studies, 16*(4), 579–604.

Clegg, S. R. (1989). *Frameworks of power*. London: SAGE.

Dawson, S. & Wedderburn, D. (1980). Introduction: Joan Woodward and the development of organization theory. In J. Woodward (Ed.), *Industrial organisation: Theory and practice* (pp. xii–xxxvi). Oxford: Oxford University Press.

Derrida, J. (1976). *Of grammatology*. Baltimore, MD: Johns Hopkins University Press.

Donaldson, L. (2001). *The contingency theory of organizations*. Thousand Oaks, CA: SAGE.

Emerson, H. (1912). *The twelve principles of efficiency*. New York: The Engineering Magazine Co.

Emery, F. E. & Thorsrud, E. (1969). *Form and content in industrial democracy: Some experiences from Norway and other European countries*. London: Tavistock.

Fayol, H. (1916). *General and industrial management*. London: Pitman.

Gergen, K. (1992). Organizational theory in the post-modern era. In M. Reed & M. Hughes (Eds.), *Rethinking organizations: New directions in organization theory and analysis* (pp. 207–226). London: SAGE.

Gillespie, R. (1991). *Manufacturing knowledge: A history of the Hawthorne experiments*. Cambridge: Cambridge University Press.

Grint, K. & Woolgar, S. (1997). *The machine at work: Nihilism and hermeneutics in post-modern culture*. Cambridge: Polity Press.

Gulick, L. (1937). Notes on a theory of organization. In L. Gulick & L. Urwick (Eds.), *Papers on the science of administration* (pp. 1–45). New York: Institute of Public Administration, Columbia University.

Hassard, J. S. (2012). Rethinking the Hawthorne studies: The western electric research in its social, political and historical context. *Human Relations, 65*(11), 1431–1461.

Hatch, M. J. & Cunliffe, A. L. (2006). *Organization theory: Modern, symbolic, and postmodern perspectives*. New York: Oxford University Press.

ITU (2019). *Statistics*. Retrieved from https://www.itu.int/en/ITU-D/Statistics/Pages/stat/default.aspx

Kelly, J. E. (1978). A reappraisal of sociotechnical systems theory. *Human Relations, 31*(12), 1069–1099.

Kijne, H. J. (1996). Time and motion study: Beyond the Taylor-Gilbreth controversy. In J. C. Spender & H. J. Kijne (Eds.), *Scientific management: Frederick Winslow Taylor's gift to the world?* (pp. 63–92). Boston, MA: Kluwer Academic Publishers.

Kjær, P. (2014). Scientific management. In S. Vikkelsø & P. Kjær (Eds.), *Klassisk og moderne organisationsteori* (pp. 47–66). Copenhagen: Hans Reitzels Forlag.

Knudsen, C. & Vikkelsø, S. (2014). Contingencyteori. In S. Vikkelsø & P. Kjær (Eds.), *Klassisk og moderne organisationsteori* (pp. 137–156). Copenhagen: Hans Reitzels Forlag.

Korver, T. (1996). Standards and the development of an internal labor market. In J. C. Spender & H. J. Kijne (Eds.), *Scientific management: Frederick Winslow Taylor's gift to the world?* (pp. 93–110). Boston, MA: Kluwer Academic Publishers.

Kreiner, K. (1992). The postmodern epoch of organization theory. *International Studies of Management & Organization, 22*(2), 37–52.

Lawrence, P. R. & Lorsch, J. W. (1967). *Organization and environment: Managing differentiation and integration.* Cambridge, MA: Division of Research, Graduate School of Business Administration, Harvard University.

Lyotard, J. (1979). *The postmodern condition: A report on knowledge.* Minneapolis, MN: University of Minnesota Press.

Lyotard, J. (1984). *The postmodern explained.* Minneapolis, MN: University of Minnesota Press.

Mayo, E. (1933). *The human problems of an industrial civilization.* New York: The Macmillan Company.

Merkle, J. A. (1980). *Management and ideology: The legacy of the international scientific management movement.* Berkeley, CA: University of California Press.

Mintzberg, H. (1980). Structure in 5's: A synthesis of the research on organization design. *Management Science, 26*(3), 322–341.

Morgan, G. (1986). *Images of organization.* London: SAGE.

Orlikowski, W. J. & Scott, S. V. (2008). Sociomateriality: Challenging the separation of technology, work and organization. *The Academy of Management Annals, 2*(1), 433–474.

Parker, M. (1992). Post-modern organizations or postmodern organization theory?. *Organization Studies, 13*(1), 1–17.

Perrow, C. (1967). A framework for the comparative analysis of organizations. *American Sociological Review, 32*(2), 194–208.

Perrow, C. (1973). The short and glorious history of organizational theory. *Organizational Dynamics, 2*(1), 3–15.

Rice, A. K. (1953). Productivity and social organization in an Indian weaving shed. *Human Relations, 6*(4), 297–329.

Roethlisberger, F. J. & Dickson, W. J. (1939). *Management and the worker.* Cambridge, MA: Harvard University Press.

Schein, E. H. (1996). Culture: The missing concept in organization studies. *Administrative Science Quarterly, 41*(2), 229–240.

Shenhav, Y. (1995). From chaos to systems: The engineering foundations of organization theory, 1879–1932. *Administrative Science Quarterly, 40*(4), 557–585.

Taylor, F. W. (1919). *Shop management.* New York: Harper. (Original work published 1903.)

Taylor, F. W. (1967). *The principles of scientific management.* New York: W. W. Norton. (Original work published 1911.)

Trist, E. L. & Bamforth, K. W. (1951). Some social and psychological consequences of the Longwall Method of coal-getting: An examination of the psychological situation and defences of a work group in relation to the social structure and technological content of the work system. *Human Relations, 4*(1), 3–38.

Vikkelsø, S. (2015). Core task and organizational reality. *Journal of Cultural Economy, 8*(4), 418–438.

Weber, M. (1922). *Wirtschaft und Gesellschaft: Grundriss der verstehenden Soziologie [Economy and society: An outline of interpretive sociology].* Tübingen: J. C. B. Mohr.

Weick, K. E. (1969). *The social psychology of organizing.* Reading, MA: Addison-Wesley.

Weick, K. E. (1990). Technology as equivoque: Sensemaking in new technologies. In P. Goodman & L. S. Sproull (Eds.), *Technology and organizations* (pp. 1–44). San Francisco, CA: Jossey-Bass.

Woodward, J. (1958). *Management and technology.* London: HMSO.

Woodward, J. (1980). *Industrial organisation: Theory and practice.* Oxford: Oxford University Press. (Original work published 1965.)

Zammuto, R. F., Griffith, T. L., Majchrzak, A., Dougherty, D. J. & Faraj, S. (2007). Information technology and the changing fabric of organization. *Organization Science, 18*(5), 749–762.

Zuboff, S. (1988). *In the age of the smart machine: The future of work and power.* New York: Basic Books.

3 PERSPECTIVES ON TECHNOLOGY AND ORGANIZATION

> In this chapter, we present three prominent perspectives on the relationship between technology and organization: technological determinism, social constructivism, and sociomateriality. We begin by considering technological determinism, the view that technological developments govern human activity. We introduce two variants of technological determinism: the pessimistic and the optimistic, as much of the literature on digital organizing implicitly or explicitly invokes one or the other normative view. We then discuss the social constructivist perspective, which is the view that the social has primacy over the technical, and that we should view technology primarily as 'texts' to be read and interpreted. Finally, we examine the sociomaterial perspective and the way in which its proponents try to strike a balance between the two other perspectives by granting equal attention to the material and the social. Toward the end of the chapter, we introduce the analytical concept of affordances, arguing that it provides a fruitful way of elucidating both material and social aspects of digital organizing.

Providing some perspective(s)

No one would dispute that organizations are impacted by the kinds of digital technology we discussed in Chapter 1. For example, online government databases allow citizens to access and manage previously hidden information, thereby reconfiguring the relationship between citizens and public servants. Big data analytics provide managers with new ways of harnessing and appropriating large quantities of data, thereby giving data a new role in decision-making. Advanced manufacturing systems allow factories to monitor and analyze work processes in real-time, thereby adding an unprecedented level of speed and precision to the production process. Social media platforms provide people with the ability to inflict lasting harm to the image of large businesses, thereby shifting the power balance between producers and consumers – and the list goes on ad infinitum.

But is it really that simple? What if these developments are not caused by digital technology as such, but by people's interpretations of or interactions with digital technology? What if the impact of digital technology is better measured by exploring how people make sense of 3D printers and robots than by trying to chart precisely how these things force humans to alter their behavior? In the final analysis, does it matter *what* actually causes organizations to change processes and procedures? Would it not be more interesting to investigate *how* these changes unfold in practice? These types of questions constitute the backbone of this chapter. In addressing such questions, we will highlight three distinct approaches: technological determinism, social constructivism, and sociomateriality.

Throughout the chapter, we refer to these as 'perspectives' on the relationship between technology and organization, in order to draw a line of association to broader discussions about the nature of reality (ontology) and our knowledge of it (epistemology). As such, each of the three perspectives corresponds to different philosophies of science. While technological determinism could be seen as a subset of the realist position within philosophy of science, social constructivism offers a contrasting view that applies 'anti-essentialist' ideas to the study of technology and organization (Grint & Woolgar, 1997). Finally, the sociomaterial perspective (Orlikowski & Scott, 2008) tries to strike a balance between the determinists and constructivists by reconciling insights from both perspectives. Toward the end of this chapter, we will explain why we consider the sociomaterial perspective to be the most appropriate strategy for studying digital organizing.

Technological determinism

Crudely put, technological determinism is the conviction that technical developments are the main factor in governing human affairs, and that these developments are driven by an internal logic resembling physical laws rather than by human activity. There are thus two corollaries to the technological determinist perspective. The first is the idea that whenever technology changes, people change their ways of interacting and organizing. Second, technology is assumed to be a factor somehow operating outside society, meaning that it evolves "independently of social, economic, and political forces" (Wyatt, 2008, p. 168). The best illustration of technological determinism in its most extreme form is perhaps the Wachowski brothers' 1999 science fiction movie *The Matrix*, which depicts a dystopian future ruthlessly governed by machines. In *The Matrix*, humans are forced to live in a simulated fantasy world that resembles present-day America, while in reality serving as bodily energy sources for the machines. Here, technology has outgrown human control and, consequently, submitted humanity to its 'will' (see Irwin, 2002).

We could provide a long list of Hollywood blockbusters that have entertained us with the story of technology running riot. Although these are obviously dramatic simplifications, similar ideas about technological dominance appear in the futuristic accounts of Artificial Intelligence and Singularity discussed in Chapter 1. Generally, however, it is rather difficult to identify organizational scholars who consider technical developments the sole determinant of human action, evolving autonomously 'outside society'. It has been argued that, today, technological determinism in its purest form has been "reduced to the status of a straw position" that allows academics to debunk certain theoretical and empirical propositions about the impact of technology on society without much intellectual scrutiny (Lynch, 2008, p. 10). Nonetheless, the idea that technical inventions shape organizational reality continues to permeate business manifestos and management rhetoric (Leonardi & Jackson, 2004), while softer versions of technological determinism are indeed detectable in academic writing. This is why we need to reclaim the notion of technological determinism and use it to "respectfully characterize works that are closer to the determinist side of the continuum of scholarly claims" instead of rejecting the perspective altogether (Dafoe, 2015, p. 1050).

The Marxist legacy

One of the first scholars to be classified as a technological determinist was the German philosopher and political economist, Karl Marx, who rose to prominence with his penetrating analysis of nineteenth-century capitalist society. As other chapters in this book will show (particularly Chapter 5), Marx is best known within scholarly circles for his magnum opus *Capital*, published during the latter part of the nineteenth century. Here, Marx (1867/1990, 1885/1992, 1894/1991) unfolds his theory of historical materialism, which holds that the mode of production (the organization of the labor process) and the relations of production that emerge from this (most notably the relationship between capital and labor) determine the organization and evolution of human society. Though Marx never suggested that technology in itself should be responsible for the process of human development, many scholars have read the theory of historical materialism as an expression of technological determinism (see Braverman, 1974, pp. 17–21). This may have something to do with this frequently cited passage:

> Social relations are closely bound with productive forces. In acquiring new productive forces men change their mode of production; and in changing their mode of production, in changing the way of earning their living, they change all their social relations. The hand-mill gives you society with the feudal lord; the steam-mill, society with the industrial capitalist. (Marx, 1847/2008, p. 119)

Here, Marx thus proposes that the emergence of new 'productive forces' is what determines the mode of production, which then ultimately shapes 'all' social relations. However, Marx's notion of productive forces referred to a wide variety of factors, ranging from raw material and labor power to machines and factories, and therefore not just to technological developments. This would make it incorrect to characterize his theory as an example of hard technological determinism. In fact, it would be more precise to portray him as an economic determinist (Bimber, 1990).

Other scholars have pulled the theory of historical materialism in directions that are more hardline (see Smith & L. Marx, 1994, for an elaboration of the distinction between hard and soft technological determinism). A prominent example of this is Robert Heilbroner's classical text 'Do machines make history?' Written on the cusp of what he calls "the age of the computer", the paper begins with the observation that technical development has obvious determining effects on the "nature of the *socioeconomic order*" (Heilbroner, 1967, p. 335). Heilbroner highlights the fact that technical inventions usually emerge in clusters (similar technologies often enter the market simultaneously), that most technical inventions are incremental rather than revolutionary ("we do not find experiments in electricity in the year 1500"), and that technical progress always seems intrinsically predictable. To support the claim that technology has determinant effects on the socioeconomic order, Heilbroner notes that the composition of the labor force seems to be directly affected by technical advances (the hand-mill requires a workforce of skilled artisans, the steam-mill requires a workforce of only semi-skilled operatives), and that the internal organization of the workplace depends on their core production technology. Based on these observations, Heilbroner (1967, p. 336) concludes:

> the steam-mill follows the hand-mill not by chance but because it is the next "stage" in a technical conquest of nature that follows one and only one grand avenue of advance. To put it differently, I believe that it is impossible to proceed to the age of the steam-mill until one has passed through the age of the hand-mill, and that in turn one cannot move to the age of the hydro-electric plant before one has mastered the steam-mill, not to the nuclear power age until one has lived through that of electricity.

Heilbroner's point about the influence of technology on the internal organization of the workplace resonates remarkably well with arguments provided by technological determinists working within the field of organization studies. Of particular interest here is the work of Joan Woodward, who we discussed in the previous chapter. To briefly recapitulate, Woodward's (1958) main point is that commercially successful corporations always modify their organizational structure to fit the complexity of their core production technology. Woodward's thesis thus reflects a rather deterministic way of thinking about the relationship between technology and organization. As Liker and colleagues (1999, p. 582) put it in summarizing her overall argument: "If you know the technology, you know the 'right' organizational form."

Other organization scholars have provided similarly deterministic accounts. One example, frequently cited alongside Joan Woodward's study of production technology and organizational structure, is Robert Blauner's (1964) investigation of working conditions in four industrial corporations. Like Woodward, Blauner argues that technology plays a key role in shaping the conditions of workers. But unlike Woodward, Blauner subscribes explicitly to Marx's vocabulary in his analysis of alienation among factory workers: "Variations in technology are of critical interest to students of the human meaning of work because technology, more than any other factor, determines the nature of the job tasks … and has an important effect on a number of aspects of alienation" (Blauner, 1964, p. 6).

Blauner's prime example of technology-dependent alienation in the workplace is the automotive industry. What makes car manufacturing stand out from other types of factory organization is its heavy reliance on assembly line technology, which tends to accentuate the division of labor and intensify worker deskilling. It takes approximately one minute for the workers to complete each of their tasks, and the level of task diversification is extremely low: "Fifty to sixty cars pass by each worker on the conveyer belt every hour, and he repeats basically the same task on a different car every minute, eight hours a day" (Blauner, 1964, p. 96). The assembly line work submits workers to the "predetermined rhythm" of the machine and deprives them of their professional autonomy (Blauner, 1964, p. 99). The powerlessness and meaninglessness experienced by workers in auto plants are contrasted with work in industries relying on craft technology (i.e. basic tools) such as printing and carpentry. Here the technology requires a high level of acquired skill and dexterity from workers, which in turn grants them significant levels of freedom on the job.

With the work of Woodward and Blauner, as well as more moderate technological determinists following their lead (e.g. Perrow, 1967; Thompson, 1967), we thus obtain an approach to the study of technology within organization studies that stresses the causal effects of technical advances on the internal organization of workplaces. Although the more moderate scholars identify some liberating effects of particularly complex technology, in the sense that it reduces managerial control and increases workers' autonomy (Liker et al., 1999), they mostly avoid a normative assessment about whether the effect of technology is positive or negative. For them, technology has direct and observable effects on organizations, but whether these effects are beneficial or harmful is an empirical question. Yet the normative discussion is never far away. Hence, in what follows, we will discuss two variants of technological determinism that are characterized precisely by their normativity in relation to technology.

Pessimism and optimism

Technology is often cast as the great *protagonist* of modern history. One need only think of the groundbreaking technical advances that occurred during the last century within fields such as aviation, medicine, chemistry, food processing, electronics, nuclear energy, space exploration, genetic engineering, and advanced manufacturing to see the instrumental role that technology has played in creating the present world. But it is precisely for this reason that technology is just as frequently cast as the great *antagonist* of modern history. Here, disastrous events like the bombing of Hiroshima and the Chernobyl nuclear accident, as well as more pervasive phenomena like global warming and ozone depletion, stand out as the most telling examples of the negative impact of technology (L. Marx, 1994). Proponents of technological optimism place much faith in technology's ability to solve our most pressing issues, whereas technological pessimists maintain that our biggest problems are caused by technology's increasing colonization of contemporary society. As Leo Marx and Merrit Roe Smith (1994, p. xii) put it:

> In the hard determinists' vision of the future, we will have technologized our ways to the point where, for better or worse, our technologies permit few alternatives to their inherent dictates. To optimists, such a future is the outcome of many free choices and the realization of the dream of progress; to pessimists, it is a product of necessity's iron hand, and it points to a totalitarian nightmare.

Technological pessimism has a long and winding pedigree, but it finds its clearest expression in the anarchist literature (Gordon, 2009). Pierre-Joseph Proudhon, the first person to officially declare himself an anarchist, was an outspoken critic of technical advances, which he perceived as "a source of wealth" for the already well-heeled and "a permanent and fatal course of misery" for the less privileged (Proudhon, 1847/2012, p. 119). For instance, while many observers saw the advent of automated production technology during the industrial era as a way of liberating workers from the drudgery of manual labor, Proudhon saw the primary purpose of industrial technology as that of driving down the cost of wages by having machines replace manual workers. This not only deprives workers of their source of living, but it also lowers the level of skills required to solve certain tasks, which is an argument that anticipates Blauner's (1964) discussion of specialization and alienation in factory work as well as Harry Braverman's (1974) work on deskilling (see Chapter 5) – and present-day discussions of robotization and losses of jobs.

Another prominent example of technological pessimism is Langdon Winner's influential writings about the (un)intended consequences of technological developments. In a paper called 'Do artifacts have politics?', Winner (1980) chronicles the story of approximately 200 bridges constructed on Long Island in New York from the 1920s and onwards by the urban developer Robert Moses. What makes these bridges stand out is their low clearance height, which allegedly prevents public busses from accessing some of Long Island's best recreational areas, as the busses are too high to pass below the crossings. According to Winner, Moses deliberately designed the bridges this way so as to ensure that only automobile-owning white people could enjoy the recreational areas, while excluding the poor and people of color. As such, the bridges effectively re-enforced racial and class-based segregation, even though the law officially prohibited this type of discrimination. Winner's point is that technologies that appear neutral or uncontroversial on first sight often serve political agendas, albeit in a concealed and subtle manner (see also Winner, 1977).

We will return to Winner's story of the Long Island bridges in the section on social constructivism. But before that, we will introduce the anti-thesis to technological pessimism, namely technological optimism. This tradition also has a long history, going back to thinkers like Henri de Saint-Simon and Thorstein Veblen, who, inspired by the spirit of Enlightenment, saw scientific and technological progress as stepping-stones on the road to emancipation. Today, the belief that technical advances will eventually solve society's most challenging problems still permeates many people's thinking. For instance, many people hope that electric airplanes will reduce carbon dioxide emissions, and in the case of digital technologies, that cryptocurrencies will increase financial transparency and that web 2.0 applications will reinvigorate democracy.

This line of thinking is clearly articulated in the work of the Russian-American author and self-proclaimed 'egoist' Ayn Rand. In a collection of essays published in response to the student uprisings of the late 1960s, Rand scolds the so-called New Left activists for hindering human progress by advocating the primacy of the collective over the individual. For Rand, technology lies at the heart of this debate. While participants in the student movements considered technology a cause of economic inequality and environmental depredation, Rand saw technological advances as the direct outcome of scientific inquiry and, hence, an expression of intellectual freedom. Human freedom would be compromised, she warned, whenever technological developments were sought curbed by way of laws or social norms. Nonetheless, Rand declares, resistance is futile, because the autonomous drive of technology cannot be stopped: "The demand to 'restrict' technology is the demand to *restrict* man's mind. It is nature – that is, reality – that makes both these goals impossible to achieve. Technology can be destroyed, and the mind can be paralyzed, but neither can be restricted" (Rand, 1971/1999, p. 285).

With the rise of digital technology, the conviction that technical progress will lead to human emancipation has gained further traction, not only in the domain of politics, but also in the academic world. A classic example is Alvin Toffler's (1980) *The Third Wave*, viewed by many observers as an illustrative example of technological optimism. As noted in the introduction to this futurist celebration of technological innovation: "The third wave is for those who think the human story, far from ending, has only just begun" (Toffler, 1980, p. 16). Toffler's core argument is that advances in particularly information technology have pushed our industrial civilization to the brink of revolution, where suddenly "humanity faces a quantum leap forward" (Toffler, 1980, p. 26). At an organizational level, Toffler predicts that these changes will usher in an upsurge in horizontal structures, flexible job commitments, and individual responsibility, thus granting workers more freedom to pursue personal interests and to adjust their work accordingly.

A number of more recent technology writers have adopted similarly optimistic positions. These include business consultants and public intellectuals such as Clay Shirky, Axel Bruns, Don Tapscott, Anthony Williams, Jeff Howe, Charles Leadbeater, and Chris Anderson, all of whom advocate a view of (digital) technology as inherently progressive. One of these digital evangelists is Kevin Kelly, editor and co-founder of *Wired* magazine. Although Kelly has written extensively about digital technology and its impact on society elsewhere, it is his 2010 monograph *What Technology Wants* that relates most clearly to the present discussion. As the title reveals, the book embarks on a quest to convince its readers that technology has an inner drive, and that the only way "we can hope to solve our personal puzzles" is by "listening to technology's story, divining its tendencies and biases, and tracing its current direction" (Kelly, 2010, p. 6).

Kelly's optimism hinges on the conviction that technological developments are driven and accelerated by a more or less autonomous sphere of life that he calls 'technicum', comprising not only concrete technologies like bicycles and iPhones, but also intangibles such as philosophical concepts and public policies (the precise limits of the technicum are not fully clear). Although technology used to be at the mercy of humans, Kelly argues, the elements making up present-day technicum have cleverly converged into a coherent system of materials and ideas, such that the technicum "often follows its own urge" – it has started to "exercise some autonomy" (Kelly, 2010, pp. 12–13).[1] It is dependent on us, but not entirely:

> The technicum wants what we design it to want and what we try to direct it to do. But in addition to those drives, the technicum has its own wants. It wants to sort itself out, to self-assemble into hierarchical levels, just like most large, deeply interconnected systems do. The technicum also wants what every living system wants: to perpetuate itself, to keep itself going. And as it grows, those inherent wants are gaining in complexity and force. (Kelly, 2010, p. 15)

As such, Kelly's argument clearly falls into the technological determinist camp, according to the way we have defined it here: (1) technology has its own logic, and (2) it evolves independent of human activity. What makes Kelly a technological *optimist* is that he views technological advances as inherently progressive. Based on the above-mentioned idea that the

---

1 Kelly frequently refers to the technicum as the "seventh kingdom of life", thus juxtaposing technology to living organisms like animals, plants, fungi, and bacteria.

technicum emulates a living organism, Kelly proposes that technology wants precisely "what life wants", which supposedly is to increase efficiency, opportunity, emergence, complexity, diversity, specialization, ubiquity, freedom, mutualism, beauty, sentience, structure, and evolvability (Kelly, 2010, p. 270). Sidestepping the questions of whether, in fact, this is what 'life' desires and how one measures things like freedom and beauty, it seems safe to conclude that most of the points on Kelly's list of what the technicum wants are positive. For instance, according to Kelly (2010, p. 309), technological inventions generally increase people's possibilities for action, which is what makes us more free and thus "more human" (see Morozov 2013, for a counter-argument). Similarly, he suggests that human beings are naturally "embedded with technophilia", meaning that everyone – regardless of aesthetic preferences – shares an inherent attraction to the beauty of technology, which is why we all apparently "love technology" (Kelly, 2010, p. 321). In the end, these assertions lead Kelly to the conclusion that the technicum "contains more goodness than anything else we know" (Kelly, 2010, p. 359), and that we should pay more attention to its deepest desires.

Few technological optimists share Kelly's somewhat bizarre views on the 'desires' and virtues of 'the technicum'. Nonetheless, the idea that the evolution of technology is unstoppable, and that we as societies and organizations should do whatever it takes to accommodate its development, is surprisingly commonplace. This is why we believe it necessary to introduce a bit of nuance to the debate on the disruptive potential of digital innovations. Unlike both optimists and pessimists, our claim is not that technology is either good or bad. Rather, we need sober thinking tools for understanding the interplay between technology and humans. In order to make that argument, we now present two additional perspectives on technology and organization: social constructivism and sociomateriality.

Social constructivism

Nearly 20 years after Winner's (1980) famous article about the Long Island bridges, Steve Woolgar and Geoff Cooper (1999) wrote a critical response. The motivation for their response was that a bus schedule published by the New York Metropolitan Transit Authority had come into the authors' possession. To their surprise, the schedule clearly showed that public busses were indeed able to pass below the otherwise low bridges, and that they do so on a daily basis. This seemed to prove beyond reasonable doubt that Winner's story about the politics of urban architecture was pure fiction – an urban legend, even. However, although they discredited the validity of a scientific paper cited more than 3,000 times, this was not Woolgar and Cooper's main intention. Instead, they proceeded to ask: To what extent does it matter if public busses are *actually* able to pass below the Long Island bridges? Because, regardless of whether it is factually correct that the bridges prevent less privileged citizens from accessing certain recreational areas, the urban legend manifests a power of its own. When such stories are produced and circulated, they help create a coherent system of meaning – a discourse – that allows people to understand the world they inhabit; and this, more than anything, is what determines how we interact and organize.

For Woolgar and Cooper (1999, p. 438), our relationship with technology is not so much dependent on the 'inherent properties' or 'internal qualities' of particular technologies, but on the discourses in which these technologies are embedded. The simple fact that Winner's story about racial and class-based segregation has proliferated so widely across time and space has contributed to establishing the idea that 'artifacts have politics', so much so that any evidence to the contrary has very limited effects. In fact, it seems safe to assume that the newly acquired bus schedule will convince few readers that Robert Moses did not originally build these bridges with malicious, if not racist, intent – and who knows if he did? There is simply no way of knowing what the 'real' politics of these bridges are. This leads Woolgar and Cooper to the conclusion that, in order for us to account for the role of technology in society, we need to understand technologies as 'texts' to be read and interpreted. Understood this way, the meaning of a given technology becomes a product of its usage rather than a result of its technical design and functionality. As they note:

> what a machine is, what it will do, what its effects will be, are the upshot of specific readings of the text rather than arising directly from the essence of an unmediated or self-explanatory technology. A technology's capacity and capability is never transparently obvious and necessarily requires some form of interpretation; technology does not speak for itself but has to be spoken for. (Grint & Woolgar, 1997, p. 32)

The notion of 'anti-essentialism' lies at the heart of the social constructivist perspective. The idea is simply that nothing – not even technology – has an inner essence; or, rather, that there is no way for us to access this inner essence. To understand the socially constructed nature of a phenomenon, including even a bridge, means we need to consider the distinction between ontology and epistemology. In the previous section, we saw that technological determinists tend to imbue technology with a certain degree of agency. For instance, Marx (1847/2008) argued that the steam-mill 'gives you' capitalist society, Jacques Ellul (1964) considered human adaptation of technology as 'caused by the action of the machine', while Kelly (2010) has gone so far as to speak of technology as equivalent to living organisms that 'want' certain things. These are all ontological claims, based on empirical observations of varying quality, which point to a number of causal connections. Through such observations, the aim for technological determinists is to expose the inner essence of technology and its relationship with society or organizations. This allows them to transport these findings across contexts and to figure out how technology in general affects organizational structures, practices, and procedures.

For technological determinists, the world exists in all its technicity independently of human thought and action. The Long Island bridges do (or do not) prevent public busses from passing below, regardless of the discourses these 'artifacts' are embedded in. The anti-essentialist spirit rejects this kind of ontologization, thus making social constructivism a purely epistemological perspective (Andersen, 2003). In its most radical form, social constructivism is not at all concerned with questions of existence (what technology is), only with questions of being (what technology means); it is concerned with our understanding of the world and not with the world ipso facto. This way of thinking resonates with much of the postmodernist literature on technology and organization that we encountered in the previous chapter such as Karl Weick's (1990) conceptualization of 'technology as equivoque', but it also ties in with broader currents within philosophy of science. Here, the work of French philosopher and historian Michel Foucault stands out as particularly relevant.

Foucault's main contribution to the social sciences has been to expose the inseparable link between power and knowledge. According to Foucault (1980), the truth (in the singular) is not something that exists 'out there' independently of human intervention. Truths (in the plural) have to be actively constructed, which means that they are subject to ongoing power-games. This does not imply an 'anything goes' approach to knowledge – far from it. However, it means that what counts as 'true' is the result of discursive struggles to promote and fortify certain understandings. The same holds for technology. There are no objective and value-free descriptions of technology's role in society. Such descriptions are always tainted by their discursive context, as well as the personal dispositions of those performing them (see Image 3.1 for an example). Diverging interests continuously struggle to consolidate particular understandings of particular technologies, which means that "we are faced with representations of technology,

Image 3.1 Alexander Graham Bell, inventor of the telephone in the 1870s, with one of the first phones. Initially, the telephone was marketed as a technology for transmitting concert music, but in time new interpretations redefined its purpose.

Source: Library of Congress/Science Photo Library, Ritzau/Scanpix.

not reflections of technology ... A reflection implies *the* truth, a representation implies *a* truth" (Grint & Woolgar, 1997, p. 33). Elaborating on the relationship between the physical world and its discursive representations, two other prominent constructivists put it like this:

> If I kick a spherical object in the street or if I kick a ball in a football match, the *physical* fact is the same, but *its meaning* is different. The object is a football only to the extent that it establishes a system of relations with other objects, and these relations are not given by the mere referential materiality of the objects, but are, rather, socially constructed. This systematic set of relations is what we call discourse. (Laclau & Mouffe, 1987, p. 82)

Several scholars have applied the social constructivist perspective to the study of digital organizing. One example is the conceptual framework developed by Rod Coombs and colleagues (1992) for analyzing applications of information and communication technology in organizations. While steering clear of what they call 'subjectivist' interpretations, they show that the discourses surrounding such technology often alter the way members of an organization understand social practices such as culture, control, and competition. They illustrate this point through a case study of the British National Health Service (NHS), where the introduction of a new budgeting program that displays the cost of medical procedures has transformed the culture of 'clinical freedom' at hospitals and in local clinics. With the new program, NHS doctors suddenly have to consider the financial cost of choosing one treatment over another, which presents an immense challenge to their professional identity. Instead of focusing purely on health issues, the doctors now have to weigh their treatment choices against the financial concerns of the healthcare system. This means that doctors are subjected to a new mode of control that works through spreadsheets rather than medical standards.

Another example of a social constructivist analysis is Todd Bridgman and Hugh Willmott's (2006) account of a large IT outsourcing contract within the field of tax administration. Drawing on the discourse theory of Ernesto Laclau and Chantal Mouffe (1985), Bridgman and Willmott argue that the effects of digital technology on different organizational setups – in this case, a partnership between the UK Inland Revenue and a private IT contractor – are wholly contingent on their discursive articulation. This means that technologies are conceived as 'social objects' that acquire a certain meaning by virtue of the context in which they are embedded. For instance, the authors show that the decision to outsource the IT contract to a private supplier was legitimated through a discourse of 'change management', in which private actors were articulated as being more knowledgeable about such processes. The change management discourse also had a visible impact on the way the technology was adopted by the tax administration, a process that had little to do with its technical design or the functional capabilities of the IT system itself.

The general point of social constructivism is not that technologies lack material existence, but that they "cannot be studied independently of their articulation in discourse" (Bridgman & Willmott, 2006, p. 120). To do so is to fall into the trap of what Keith Grint and Steve Woolgar (1997) call 'technicism', that is, attributing to technology certain undeniable effects. Social constructivism thus represents the antithesis to technological determinism. It contests the technological determinists' assumption of a monocausal relationship between the inherent properties of technology and organizational realities.

Is there some way to strike a balance between the two perspectives? Is there some way to maintain a constructivist-inspired epistemology (i.e. knowledge is the product of ongoing power-games), while acknowledging that technologies constrain and enable certain actions, regardless of the wider discursive context? We believe that this compromise lies with our third approach, sociomateriality. The sociomaterial approach was initially developed under various headings, including 'the social construction of technology' (e.g. Bijker et al., 1987) and 'the social shaping of technology' (e.g. MacKenzie & Wajcman, 1985). In this book, however, we retain the more recent label – sociomateriality. Our goal here will be to show how sociomateriality differs from the radically anti-essentialist spirit of social constructivism by emphasizing the role played by materiality.

Sociomateriality

It is commonly argued that studies of digital technologies suffer from a deterministic slant *or* that they fail to acknowledge the role of technology. In organization studies, the polarity of these positions has become the subject of much discussion, leading some scholars to look for inspiration within Science and Technology Studies (STS) to address the relationship between technology and organizing processes in a less dichotomous way. The Sociology of Science, a subfield of sociology, emerged with an interest in explaining how science is not just driven by objective and value-free knowledge interests, but depends on the social structures in which it is embedded. At a certain point, some STS scholars began questioning the relative significance accorded to the *social* context and *social* constructions influencing the production of science – and the relative neglect of the material aspects of science. They argued for the need to cultivate a more 'symmetrical' approach to social phenomena, taking into account how not only humans but also 'non-humans' influence them (e.g. Callon, 1986).

Today, this 'post-humanist' approach has become more of a general sociological theory. One of the founding fathers of STS, Bruno Latour (2005), has developed an extensive critique of what he calls 'the sociology of the social'. Latour claims that 'the social' does not exist in and of itself; it cannot be identified. We are left to speculate and theorize about 'social phenomena', simply because we cannot observe them. As an alternative, Latour launches a 'sociology of associations'. Rather than trying to theorize how hidden social structures govern relationships or observable phenomena, Latour promotes the idea of simply observing what happens in practice; how relations are made and by what means they are sustained. The central point here is that relationships are often established between human and non-human actors, and that the non-human (material, technological, linguistic, bodily, etc.) elements are often very important parts of these relationships (Latour, 1992). Such ideas eventually evolved into what is known as actor-network theory (ANT).

John Law (1997), another proponent of this sociology of associations and a major figure in ANT, beautifully illustrates the importance of materiality in an anecdote about a study of management. Although management is conventionally considered a purely social phenomenon, Law finds management to be intrinsically bound up with material elements. He describes how, in his capacity as researcher, he once entered a glass door, climbed a set of impressive stairs, and entered the office of a top manager. In that particular situation, the manager was furious and displayed his power effectively, making Law wonder where all that power came from. Some of it was clearly linked to symbols like the grand conference table, the phones, the secretary, and so on. But besides this, Law makes a thought experiment in his account, introducing a deconstructive fairy with evil intentions. This evil fairy takes away the manager's computer. The computer turns out to be more than an electronic box with a keyboard. It is also the manager's spreadsheet, his budgets, and his projections. All of a sudden, the super powerful manager can no longer calculate. He has no idea about the financial situation of the organization. Without a word processor, he can no longer write as easily. What is worse, however, is that his archives, publications, and electronic calendar are also gone, and the manager is now a man without a past and a memory. On top of this, he has no phone, email, or fax – and thus no contacts. As Law puts it somewhat polemically, when a man can no longer calculate, write, or remember anything, he is no longer a powerful person, but rather "a naked ape – with all the powers of a naked ape!" (Law, 1997, p. 3).

Law's story is not just about the manager needing material elements to be a manager. There is also another side to the manager's practice: all the material elements force the manager to act in particular ways. Associations between accountants, computers, and spreadsheets have created files, and then pieces of paper, that force him to (re)act in particular ways. As such, the sociomaterial relations within the organization create 'the manager' (as an occupational role). Law's story is thus an attempt to illustrate not only that people (like managers) have agency, but also that seemingly social phenomena (like management) are thoroughly entangled with material elements (see also Law & Hassard, 1999). This way of thinking has inspired a number of organization and management scholars. One important translator of the 'symmetrical approach' to humans and non-humans into organization studies is Wanda Orlikowski. She originally introduced the term 'socio-materiality' to highlight the inseparability of sociality and materiality in organizations and thus challenge the traditional separation between technology, work, and organization. In later work, she even omitted the hyphen between 'socio' and 'material', so as to underscore the constitutive entanglement of the social and the material (Orlikowski & Scott, 2008).

In a key article, Orlikowski (2007) offers a systematic critique of organization studies, claiming that the discipline as a whole tends to disregard, downplay, or take for granted materiality in organizations. She also points out that many studies dealing with technology in organizations are focused on specific cases of technology implementation or use, thus treating technology as some kind of special case. She calls the first approach 'human-centered' and the second approach 'techno-centric'. The human-centered approach focuses on how humans make sense of and interact with technology. Here, technology is black-boxed and its role minimized. The techno-centric view tends to assume that technology is homogeneous, predictable, and stable. In this view, focus remains on the humans who interact with a more or less 'dead' technology. In contrast to these two positions, Orlikowski proposes that we consider materiality integral to organizing. We need, she says, to think of organizational practices as *sociomaterial* rather than just *social* practices.

Orlikowski exemplifies her sociomaterial approach with analyses of two digital technologies. The first technology is the Google search engine. Here, Orlikowski shows how researchers' everyday work practices are entangled with the materiality of the 'Page Rank' algorithm that 'decides' which pages are displayed highest up on the list when we search. She argues that the displays are as dependent on the searches as the search-outcomes are dependent on the displays. Together, humans and technology produce a thoroughly sociomaterial research practice. The second example of sociomaterial technology is the BlackBerry. In this particular case, Orlikowski argues that mobile communication changes the entire way in which organizational members interact: not just the how, but the when, where, and even the why. The 'push email' capability of the BlackBerry's software is entangled with changing norms of communication in a company where employees become increasingly addicted to being in the loop. Together, the sociomaterial practices result in a blurring of the boundary between work and the private sphere. The sociomaterial perspective has become increasingly popular in analyses of digital organizing, but has not replaced the other two perspectives. All three can be discerned in contemporary work (see Table 3.1 for an overview).

A commonly used analytical concept associated with the sociomaterial perspective is that of 'affordances'. The concept of affordances has been most frequently deployed by researchers working within the sociomaterial tradition (e.g. Carlile et al., 2013; Leonardi et al., 2012), but all the perspectives we have presented here can usefully be discussed up against this

Table 3.1 Overview of the three perspectives on technology and organization.

	Technological determinism	Social constructivism	Sociomateriality
Philosophy of science	Realism / empiricism	Social constructivism	(Social) constructivism
View of technology	Demarcated entities with intrinsic features that cannot be denied	Texts being written (articulated) and read (interpreted)	Actors influencing and being influenced by other actors
Analytic focus	Causal relationships between technology and human activity	Discursive representations of technology and its impact of human activity	Technology's role in shaping networks of heterogeneous actors
Notable contributions	Marx, 1847 Blauner, 1964 Woodward, 1965 Heilbroner, 1967 Winner, 1980 Toffler, 1980	Foucault, 1980 Weick, 1990 Coombs et al., 1992 Grint & Woolgar, 1997 Woolgar & Cooper, 1999 Bridgman & Willmott, 2006	Callon, 1986 Latour, 1992 Law & Hassard, 1999 Latour, 2005 Orlikowski & Scott, 2008 Leonardi et al., 2012

concept, to clarify one's analytical focus. Let us begin, however, by describing the conceptual genealogy of affordances.

The concept of affordances

The concept of affordances was originally coined by the psychologist James Gibson (1979). It seeks to account for how animals perceive and interact with the natural world. Gibson's most basic point is that animals (including humans) perceive their environment in terms of usage, and that the utility of something is at least partially constituted in the eyes of the beholder. The concept of affordances describes precisely this point. When Gibson authored his seminal work on perception, the dictionary contained an entry on the verb 'to afford', meaning 'allowing for something'. However, the noun 'affordance' did not exist. Gibson thus coined the term to refer both to the properties of a material environment and to the individual's perception and use of that environment. The affordance concept expresses the idea that our material surroundings *frame* our possibilities for action, but do not *determine* them. Hence, our abilities to act in relation to our surroundings depend on a combination of our perception, the cultural context in which we find ourselves, and the properties of the material elements we seek to use.

In its purest form, this view of technology is neither social constructivist nor determinist. A social constructivist view would imply that we focus on technologies as 'texts' to be read and interpreted in a social setting, whereas a determinist view would imply that technologies come with the inanimate power to alter the social. For this reason, the inspiration from the affordance concept might help us avoid the kind of techno-optimist, disruption hype described in Chapter 1. For instance, media scholar Ian Hutchby has argued that an affordance approach helps to distinguish his work on communication technologies from "more hyped-up journalistic versions of 'technological revolution', 'depersonalized communication' and 'information overload' that in themselves simply echo, under a new guise, technophobic and technophiliac concerns" (Hutchby, 2001, p. 195).

The affordance concept has been used in organization studies to create practice-based analyses of various artifacts and technologies; the goal is to show how otherwise 'social phenomena' are intertwined with non-human elements. For instance, Sarah Kaplan (2011) analyzes the affordances of PowerPoint presentations in connection with strategy work. Kaplan describes how PowerPoint technology mobilizes knowledge production in a specific way. She argues that PowerPoints have several affordances that shape collaboration around strategy making. First, they offer materiality to strategic ideas by displaying what is not yet a reality. At the same time, the PowerPoint slides are mutable and modular. They can be edited, deleted, or moved around through the collaboration process. They share these properties with other textual and digital technologies, but Kaplan observes how people *related to* (or, to use Gibson's terminology, *perceived*) the PowerPoints in a rather special way: "Instead of being asked to do a new analysis, a team member would be asked to provide a slide on the topic; instead of disagreeing about an idea, participants disagreed with "charts"; deliverables were described in terms of "chart decks" or "packs" rather than in terms of strategies or decisions" (Kaplan, 2011, p. 327). In this way, PowerPoints were used not just to present ideas, but afforded a particular kind of collaboration: "Progress was measured in slides. Time was measured in slides. Strategic discussions could not take place if the slides to support them were not available or correctly formatted" (Kaplan, 2011, p. 327).

Organization scholars have also analyzed how social media affords new organizational practices. For instance, Jeffrey Treem and Paul Leonardi (2012) argue that the affordance concept allows us to understand the organizational consequences of social media software. They criticize previous studies of the role of social media in organizations for either just pointing to their properties (that they are web-based or allowing for user-generated content) or exemplifying what social media are (e.g. blogs or wikis). This leads to studies that limit their focus to a particular tool in a particular organizational context, thus preventing us from theorizing social media use in organizations at a more abstract level. Treem and Leonardi suggest that we focus on the kinds of behaviors typically afforded by social media across organizations, as this will allow us to understand when, why, and how social media lead to changes in organizational practices.

By reading an extensive literature on social media use in organizations, the authors observe that social media seem to have four different affordances that potentially change organizational processes. Social media afford a high degree of visibility, persistence, editability, and association. The 'visibility' affordance implies that users of social media can make elements of their behaviors, knowledge, and network connections – which were once hidden – visible to other organizational members. The 'persistence' affordance points to the fact that communication is stored and can be used as collective memory. The 'editability' affordance is about asynchronous and isolated communication; social media users have ample opportunity to craft their contributions. Finally, 'association' implies that not only are individuals tied together by social media, but they are also tied to various pieces of information, implying that the information provided in social media networks is never impersonal. Treem and Leonardi suggest that these four affordances may lead to changes in organizational processes of socialization, in information sharing, and in power relations – and they urge researchers to find out exactly how.

Raymond Zammuto and colleagues (2007) also use the concept of affordances to capture possible changes in organizing. They emphasize that affordances of digital technologies do not depend solely on the properties of these technologies, but also on the organization's capacities. Only when we understand both technological and organizational features – and their interconnectedness – will we be able to reflect on changes in organizational form and function. The authors offer five examples of instances where digital technologies and organizational design are woven together in such a way as to afford particular organizational processes. They call the first affordance 'visualization of entire work processes' and then go on to discuss how technological ensembles of databases, dashboards, sensors, and particular software allow for the monitoring of work processes when these technologies are coupled with process-centered organizing. This then affords easy identification of interdependencies in the organization, as well as understanding of the role played by different actors in organizational processes. Zammuto and colleagues analyze four other affordances in a similar manner. They zoom in on real-time product and service innovation, virtual collaboration, mass collaboration, and simulation features, arguing that these affordances create new organizational forms.

From these examples, it could seem like the notion of affordances is very much about revealing the intrinsic potentials of technologies – that is, the possibilities offered by various designs and functionalities. However, in order to use the concept as intended, we must hold on to the idea that affordances are also a product of the user's perception of digital technologies, as well as the cultural context surrounding the technology and its user(s). This is emphasized in an affordance analysis of big data, where Anders Koed

Madsen (2015) argues against the determinist view that digital data 'speak for themselves'. Madsen highlights how perception plays a major role in constructing knowledge on the basis of seemingly neutral data.

Analytical focus

We will use the remainder of this chapter to suggest that all three perspectives can fruitfully be understood in relation to three dimensions of the affordance concept. As described above, Gibson (1979) initially coined the term 'affordances' to escape the dualism embedded in the objectivist/subjectivist relationship. As such, the concept of affordances was originally meant to help researchers consider how material environments and subjective interpretations co-constitute certain possibilities for action. Later, Hutchby (2001) added a third dimension to the theory, namely social norms and conventions governing the use of a certain object (see also Kamstrup & Husted, forthcoming). Conducting an affordance analysis is thus a matter of considering:

1) The material properties of a given object (its design, functionality, aesthetics, etc.)
2) The subjective interpretations of a given object (how people make sense of it)
3) The social context surrounding a given object (rules, norms, conventions, etc.).

Combining these three dimensions allows us to probe the affordances of a certain object or environment – that is, it brings us to a position where we can make qualified assessments regarding "the possibilities for action brought forth by it to a perceiving subject" (Fayard & Weeks, 2007, p. 609). This triadic way of thinking about the affordances of something is obviously somewhat artificial. After all, how do we know when individual perception stops and social norms take over? Moreover, how can we describe the properties of particular technologies without using our own subjective perception and without drawing on cultural norms embedded in our own social context? The answers to these questions are, of course, that we do not and cannot. This means that the three dimensions must be thought of as focal points. They help draw our attention to certain aspects of a given technology, but without claiming that the three aspects can be separated empirically and described objectively.

Framing the triad as a matter of focus provides us with a certain degree of analytical freedom. First, a triad approach allows us to embrace the empirical messiness of digital organizing instead of trying to compartmentalize the social world into neatly demarcated boxes. Second, it opens up the possibility of shifting focus from one area to another. If we position ourselves as sociomaterialists, we try to strike a balance between all three dimensions by giving equal attention to materiality, sociality, and individual perception. If,

DIGITAL ORGANIZING

Socio-materiality	Social Constructivism	Techno Determinism
Material properties	Material properties	**Material properties**
Subjective interpretations	**Subjective interpretations**	Subjective interpretations
The social context	**The social context**	The social context

(Affordances)

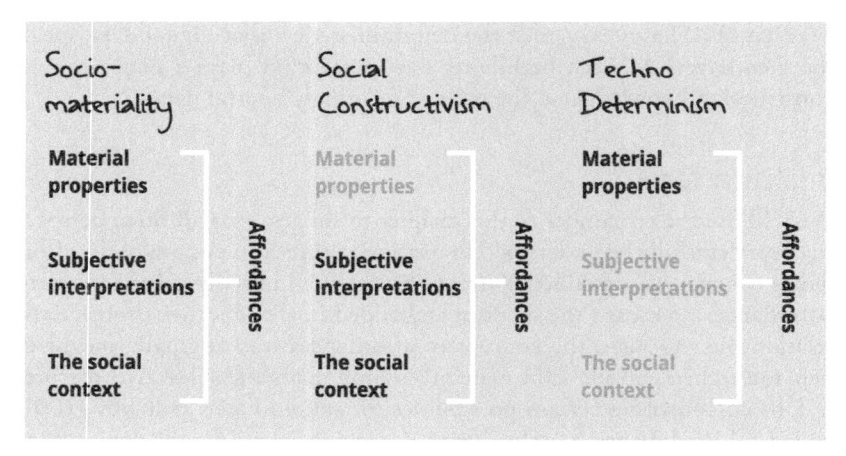

Figure 3.1 Aspects of the affordance concept. The sociomateriality perspective addresses all three aspects. The two other perspectives mostly address one or two aspects.

on the other hand, we position ourselves as technological determinists, we grant priority to the material side of the triad by focusing predominantly on the design and functionality of a given technology. And finally, if we position ourselves as social constructivists, we downplay materiality and focus more exclusively on the social context and individual readings of the technology in question. Figure 3.1 illustrates this line of reasoning.

Implications for methods and analysis

Let us consider an example. Assuming we want to conduct an analysis of the possibilities for action afforded by social media in relation to user-involvement in online news production, we could strive to obtain information about all three dimensions. First, we could try to find out how platforms like Facebook and Twitter operate at the technical level: What functions do they offer users, and how are they geared algorithmically? Second, we could assemble information about how individual users understand the relationship between news production and social media: Do they only use the platforms to consume and circulate news, or are they actively engaged in news production by uploading pictures or writing comments on the pages of news agencies? Finally, we could explore the social context engulfing platforms like Facebook and Twitter: In what kind of discourses are these platforms embedded, and how do these discourses impose certain norms on their users in relation to online news production and exclude others?

Depending on one's perspective, parts of this information vary in importance. Technological determinists would emphasize the first dimension, for instance by investigating the complex infrastructure of hashtag networks, and how the algorithmic ordering of newsfeeds tends to create filter bubbles. Social constructivists would prefer to focus on the second and third dimensions, for instance unpacking popular discourses of 'co-creation' and 'crowdsourcing' and exploring users' willingness to identify with these narratives. Sociomaterialists, meanwhile, would be searching for a way to merge these distinct interests by focusing on what actually goes on at the platforms in relation to news production. This brief example illustrates that the three perspectives navigate according to rather different analytical foci. In the example here, technological determinists focus on technical *designs*, social constructivists focus on systems of *meaning*, and sociomaterialists focus on emerging *practices*.

Obviously, these differences have methodological implications. It is rather difficult to figure out how Google's Page Rank algorithm works by interviewing ordinary internet users, just as it is impossible to understand how people assign meaning to the Cambridge Analytica scandal by reading design manuals. We will not go into detail with considerations about methodology, but merely point out that the choice of perspective has a direct bearing on the methods applied to understand digital organizing. To understand the material properties of a given technology, one needs data sources that provide this type of information, which would typically involve policy documents, technical documentation, statistics, and possibly expert statements. To understand the systems of meaning that govern the use of a particular technology, one needs data sources that open a window to such information. This kind of data is usually derived from interviews, focus groups, and reading documents of relevance to the discourse one is trying to analyze. To understand how certain practices emerge from the intersection of technology and sociality, one needs methods that illuminate this particular area of interest. The preferred method here would often be ethnographic observations, but big data visualizations and network analyses could also be options (for more comprehensive examinations of methods in relation to digital organizing, see Hine, 2015; Kozinets, 2010; Paulus et al., 2013; Pink, 2013; Rogers, 2013).

Conclusions

In this chapter, we have provided an overview of three prominent perspectives on the relationship between technology and organization: technological determinism, social constructivism, and sociomateriality. We have used a relatively large brush to paint this picture, emphasizing differences between these approaches while showing the analytical potentials of each. In focusing on the differences between the three perspectives, they end up appearing more dogmatic than they truly are. After all, as noted by Grint & Woolgar (1997), very few scholars are willing to commit 100% to either side of the determinist/constructivist debate (although their own writings represent one of the clearest expressions of the social constructivist stance). For this reason, perhaps, the sociomaterial perspective has gained considerable traction among scholars. Not only does it seek to resolve the dichotomy between constructivism and determinism, but it typically also yields more nuanced analyses.

We are inspired by sociomateriality in our approach to the themes of the following chapters. This does not mean that our accounts of, respectively, structure and infrastructure, production and produsage, collaboration and co-creation, knowledge and datafication, communication and interactivity, legitimacy and transparency, power and empowerment are based only on sociomaterial studies. In fact, we rely on a broad range of both technological determinist and social constructivist sources as well. Rather, we feel that the sociomaterial perspective is an effective way for us to bring nuance to the technology-organization discussion. As such, each of the following seven chapters concludes by discussing the complementarity of classical and contemporary concepts from organization studies with a sociomaterial sensibility. This allows us to see that there is indeed something valuable about the many buzzwords that constantly enter our digital parlance, because they point to ways in which organizations are *potentially* affected by the emergence of new digital technologies. However, unreflective use of these buzzwords blinds us to the ways in which organizations may in fact *not* change when subjected to technological developments. For instance, if we substitute the old notion of power with the fashionable idea of empowerment, we fail to see that these concepts are not necessarily antinomic, and that digital technology cannot help us escape the grip of power. Demonstrating some of the continuities in the dynamics of organizations is crucial if we want to understand digital organizing in the twenty-first century.

? QUESTIONS

1. To what extent can we speak of digital technology evolving autonomously and driving societal changes?
2. Do some technologies seem to have an inner essence and, if so, how do we study this essence?
3. How is a given digital technology shaped by its discursive context?
4. What types of discourses surround given contemporary digital technologies?
5. How and when does it make sense to speak of digital technology as an 'actor' that does something?
6. What kind of limits are there to such an 'agency' perspective when applied to digital technology?
7. Which organizational practices are afforded by digital technologies like social media or robotics?

References

Andersen, N. Å. (2003). *Discursive analytical strategies: Understanding Foucault, Koselleck, Laclau, Luhmann.* Bristol: Policy Press.

Bijker, W. E., Hughes, T. P. & Pinch, T. (1987). *The social construction of technological systems: New directions in the sociology and history of technology.* Cambridge, MA: MIT Press.

Bimber, B. (1990). Karl Marx and the three faces of technological determinism. *Social Studies of Science, 20*(2), 333–351.

Blauner, R. (1964). *Alienation and freedom: The factory worker and his industry.* Chicago, IL: University of Chicago Press.

Braverman, H. (1974). *Labor and monopoly capital: The degradation of work in the twentieth century.* New York: Monthly Review Press.

Bridgman, T. & Willmott, H. (2006). Institutions and Technology: Frameworks for understanding organizational change: The case of a major ICT outsourcing contract. *The Journal of Applied Behavioral Science, 42*(1), 110–126.

Callon, M. (1986). Some elements of a sociology of translation: Domestication of the scallops and the fishermen of St Brieuc Bay. In J. Law (Ed.), *Power, action and belief: A new sociology of knowledge?* (pp. 196–223). London: Routledge & Kegan Paul.

Carlile, P., Nicolini, D., Langley, A. & Tsoukas, H. (2013). *How matter matters: Objects, artifacts, and materiality in organization studies.* Oxford: Oxford University Press.

Coombs, R., Knights, D. & Willmott, H. C. (1992). Culture, control and competition: Towards a conceptual framework for the study of information technology in organizations. *Organization Studies, 13*(1), 51–72.

Dafoe, A. (2015). On technological determinism: A typology, scope conditions, and a mechanism. *Science, Technology, & Human Values, 40*(6), 1047–1076.

Ellul, J. (1964). *The technological society.* New York: Alfred A. Knopf.

Fayard, A.-L. & Weeks, J. (2007). Photocopiers and water-coolers: The affordances of informal interaction. *Organization Studies, 28*(5), 605–634.

Foucault, M. (1980). *Power/knowledge: Selected interviews and other writings, 1972-1977.* New York: Pantheon.

Gibson, J. (1979). *The ecological approach to visual perception.* Hillsdale, NJ: Lawrence Erlbaum Associates.

Gordon, U. (2009). Anarchism and the politics of technology. *The Journal of Labor and Society, 12*(3), 489–503.

Grint, K. & Woolgar, S. (1997). *The machine at work: Technology, work and organization.* Cambridge: Polity Press.

Heilbroner, R. L. (1967). Do machines make history? *Technology and Culture, 8*(3), 335–345.

Hine, C. (2015). *Ethnography for the internet: Embedded, embodied, everyday.* London: Bloomsbury Academic.

Hutchby, I. (2001). *Conversation and technology: From the telephone to the Internet.* Cambridge: Polity Press.

Irwin, W. (2002). *The Matrix and philosophy: Welcome to the desert of the real.* Chicago, IL: Carus Publishing.

Kamstrup, A. & Husted, E. (forthcoming). Is crowdsourcing anarchist? Interrogating the libertarian spirit of two online platforms. In M. Parker, T. Swann & K. Stoborod (Eds.), *Anarchism, organization and management.* London: Routledge.

Kaplan, S. (2011). Strategy and PowerPoint: An inquiry into the Epistemic Culture and Machinery of Strategy Making. *Organization Science*, *22*(2), 320–346.
Kelly, K. (2010). *What technology wants*. New York: Viking Books.
Kozinets, R. (2010). *Netnography: Doing ethnographic research online*. London: SAGE.
Laclau, E. & Mouffe, C. (1985). *Hegemony and socialist strategy: Towards a radical democratic politics*. London: Verso Books.
Laclau, E. & Mouffe, C. (1987). Post-Marxism without apologies. *New Left Review*, *166*(1), 79–106.
Latour, B. (1992). Where are the missing masses? The sociology of a few mundane artifacts. In W. Bijker & J. Law (Eds.), *Shaping technology – Building society* (pp. 225–259). Cambridge, MA: MIT Press.
Latour, B. (2005). *Reassembling the social: An introduction to actor-network-theory*. Oxford: Oxford University Press.
Law, J. (1997). *The manager and his powers*. Lancaster: Centre for Science Studies, Lancaster University. Retrieved from http://www.lancaster.ac.uk/fass/resources/sociology-online-papers/papers/law-manager-and-his-powers.pdf
Law, J. & Hassard, J. (1999). *Actor network theory and after*. Oxford: Blackwell.
Leonardi, P. M. & Jackson, M. H. (2004). Technological determinism and discursive closure in organizational mergers. *Journal of Organizational Change Management*, *17*(6), 615–631.
Leonardi, P. M., Nardi, B. A. & Kallinikos, J. (2012). *Materiality and organizing: Social interaction in a technological world*. Oxford: Oxford University Press.
Liker, J. K., Haddad, C. J. & Karlin, J. (1999). Perspectives on technology and work organization. *Annual Review of Sociology*, *25*, 575–596.
Lynch, M. (2008). Ideas and perspectives. In E. Hackett, O. Amsterdamska, M. Lynch & J. Wajcman (Eds.), *The handbook of science and technology studies* (pp. 9–12). Cambridge, MA: MIT Press.
MacKenzie, D. & Wajcman, J. (1985). *The social shaping of technology*. London: Open University Press.
Madsen, A. K. (2015). Between technical features and analytic capabilities: Charting a relational affordance space for digital social analytics. *Big Data & Society*, *2*(1), 1–15.
Marx, K. (1990). *Capital, volume 1*. London: Penguin Books. (Original work published 1867.)
Marx, K. (1991). *Capital, volume 3*. London: Penguin Books. (Original work published 1894.)
Marx, K. (1992). *Capital, volume 2*. London: Penguin Books. (Original work published 1885.)
Marx, K. (2008). *The poverty of philosophy*. New York: Cosimo Classics. (Original work published 1847.)
Marx, L. (1994). The idea of 'technology' and postmodern pessimism. In L. Marx & M. R. Smith (Eds.), *Does technology drive history?* (pp. 237–258). Cambridge, MA: MIT Press.
Marx, L. & Smith, M. R. (1994). *Does technology drive history? The dilemma of technological determinism*. Cambridge, MA: MIT Press.
Morozov, E. (2013). *To save everything, click here: Technology, solutionism and the urge to fix problems that don't exist*. London: Allen Lane.
Orlikowski, W. J. (2007). Sociomaterial practices: Exploring technology at work. *Organization Studies*, *28*(9), 1435–1448.

Orlikowski, W. J. & Scott, S. V. (2008). Sociomateriality: Challenging the separation of technology, work and organization. *The Academy of Management Annals, 2*(1), 433–474.

Paulus, T., Lester, J. & Dempster, P. (2013). *Digital tools for qualitative research*. London: SAGE.

Perrow, C. (1967). A framework for the comparative analysis of organizations. *American Sociological Review, 32*(2), 194–208.

Pink, S. (2013). *Doing visual ethnography*. London: SAGE.

Proudhon, P. J. (2012). *System of economical contradictions: Or, the philosophy of poverty*. Portland, OR: The Floating Press. (Original work published 1847.)

Rand, A. (1999). *The return of the primitive: The anti-industrial revolution*. New York: Meridian. (Original work published 1971.)

Rogers, R. (2013). *Digital methods*. Cambridge, MA: MIT Press.

Thompson, J. D. (1967). *Organizations in action*. New York: McGraw-Hill.

Toffler, A. (1980). *The third wave*. London: Pan Books.

Treem, J. W. & Leonardi, P. M. (2012). Social media use in organizations: Exploring the affordances of visibility, editability, persistence, and association. *Communication Yearbook, 36*, 143–189.

Weick, K. E. (1990). Technology as equivoque: Sensemaking in new technologies. In P. Goodman & L. Sproull (Eds.), *Technology and organizations* (pp. 1–44). San Francisco, CA: Jossey-Bass Publishers.

Winner, L. (1977). *Autonomous technology: Technics-out-of-control as a theme in political thought*. Cambridge, MA: MIT Press.

Winner, L. (1980). Do artifacts have politics? *Daedalus, 109*(1), 121–136.

Woodward, J. (1958). *Management and technology*. London: HMSO.

Woolgar, S. & Cooper, G. (1999). Do artefacts have ambivalence? Moses' bridges, Winner's bridges and other urban legends in S&TS. *Social Studies of Science, 29*(3), 433–449.

Wyatt, S. (2008). Technological determinism is dead; long live technological determinism. In E. Hackett, O. Amsterdamska, M. Lynch & J. Wajcman (Eds.), *The handbook of science and technology studies* (pp. 165–180). Cambridge, MA: MIT Press.

Zammuto, R. F., Griffith, T. L., Majchrzak, A., Dougherty, D. J. & Faraj, S. (2007). Information technology and the changing fabric of organization. *Organization Science, 18*(5), 749–762.

PART II

REVISITING THEMES IN ORGANIZATION STUDIES

4 STRUCTURE AND INFRASTRUCTURE

> This chapter highlights the role of organizational structures in the way organizations achieve collaboration, communication, and control. These organizational structures can be both formal and informal, they are frequently the object of redesign, and they can be depicted in diagrams to illustrate connections and division of labor. While organizational structures continue to be of practical concern and an object of organizational studies, scholars have argued they also have a tendency to become static and out of sync with a more dynamic and messy organizational reality. This chapter discusses such critiques, focusing on the networked organization and the virtual organization as alternative forms of organizing. Alternative forms of organization emerging with the digitalization of organizations may require a different vocabulary to describe their operations, which has led us to cultivate the concept of infrastructure. The infrastructure concept can take account of some of the technological, material, and symbolic elements that are central to organizing. This does not mean that in analyzing digital organizing, the concept of structure has become obsolete. It means only that the concept of infrastructure has some additional explanatory power that can grasp those organizational phenomena that are important to digital organizing.

Why structure and infrastructure?

Classical organization theory was concerned with cooperative systems and administrative principles (e.g. Taylor, 1911/1967; Gulick, 1937/2005; Barnard, 1938). This early work has since been interpreted as having 'organizational structures' as a key concern. Organizations have been conceptualized as complex wholes, such that the principal task for the specialist (manager or researcher) was to analyze or plan how the parts of the whole would work together to pursue the organization's goal (Du Gay & Vikkelsø, 2017). Questions of authority, division of labor, and coordination were thus addressed by analyzing or planning the formal structuring of the organization, along with paying attention to informal structures (Blau & Scott, 1962; Mayo, 1933; Selznick, 1948). Since then, organizational structures have been used to explain many different organizational issues. For instance, it has been proposed that an organization's interaction with its public depends on its formal structures. In 1962, Peter Blau and W. Richard Scott wrote that:

> Which level in the organizational hierarchy deals with the major public-in-contact varies greatly. In unions, for example, it is the top-level: leaders negotiate with management. But in prisons, it is the bottom level: guards have direct contact with inmates. Two main factors appear to determine the hierarchical level from which are selected the organization's representatives who work with a given public. The smaller the number of these representatives (compared to the entire organization), and the higher the status and power of the public with whom they deal, the higher will be their relative standing in the organization. (Blau & Scott, 1962, p. 60)

This way of thinking about the relationship between an organization and its environment is informed by an interest in levels, hierarchies, and division of labor – in other words, organizational structures. In a quote like the one above, it is assumed that particular organizations have particular traits and structures, and that these traits and structures determine various aspects of behavior. More recent organizational studies have shown that digital technologies tend to disturb organizational and occupational structures by, for instance, affecting the relations between different occupations (e.g. Barley, 1986). One might therefore ask what happens when an organization has multiple (digital) platforms for interaction? Most would agree that the organization's interaction with its environment and other organizational phenomena will depend on many more factors than formal structures. Moreover, formal structures might be put under pressure by, for instance, digital infrastructures.

Structures and infrastructures: The case of a virtual library

To remain with the example of interaction with the environment, we see indications that organizational structures are indeed subjected to challenges by digital technologies. A story of the digitalization of a library offers a telling example of how digital organizing alters an organization's relations with its users as well as its hierarchical structures. The story goes that a university chose to turn a part of its library into a book-free institution. As the library went virtual, it became difficult for librarians to uphold the kind of relations they had with their students and researchers previously, when they served as 'custodians of the books'. When books were physical objects, librarians were also visible as someone to consult, and their role in pursuing the purpose of the organization was clear. As the organization was turned into an environment where information search happened on computers only, the librarians were no longer recognized as relevant professionals to consult, and their place and role in the organization changed.

Organizational scholars observed how the librarians felt compelled to work with redefining their tasks and professional identities when all material was digitized (Boudreau et al., 2014). Since they were no longer approached by users, they began to advertise their presence and the possibility of asking for help. However, these efforts resulted in a large number of directional or practical questions. Frustrated, the librarians were not interested in becoming clerical workers (which would amount to a change of tasks and a loss of status), so they decided to try to approach citizens online and put their professionalism to work there.

It can be argued that, in this example, no *formal* organizational structures were changed. Formally, the librarians still occupied the same position in the organization, and they were still supposed to serve the same public. Furthermore, their main task – assisting with information searches – had not been formally changed. The big change was the digitization of formerly physical resources. However, the interpretation of digital technologies and the work around them entailed unintended changes of tasks and ongoing work by the librarians to redefine their professional role vis-à-vis the public. It can thus be argued that to understand the library organization, it is important to not only look at formal structures, but to also take account of the less visible *digital infrastructures* that may pose a challenge to formal structures. The term 'infrastructure' points to some of the material or technological underpinnings of an organization; these are elements that are not covered by the term 'organizational structure'. 'Infra' means 'below', and infrastructures are important because they have effects even though they normally operate under the radar. As Geoffrey C. Bowker and Susan Leigh Star (1999) write, infrastructures become visible only when they break down. Or, as we saw in the case of the virtual library, when the infrastructure is remade.

It would be exaggerated, however, to say that digital infrastructures simply replace organizational structures as relevant objects of study. Both need to be examined if we are to understand some of the basic features that create or sustain an organization. In the first part of this chapter, we will look back at what organizational theory has to offer in terms of understanding organizational structures. We will then explore how a focus on digital infrastructures may help us better understand aspects of digitalized organizations. Let us first begin with the concept of 'structure' itself.

Structure as a theme in organization studies

To create or reflect on organizational structures is basically a task of determining who does what in an organization and how different people or entities relate to one another. This makes organizational structure a very practical concern. Structure has also been a focus of organizational scholars, who have been interested in how choices about structures can be linked to organizational success. A very influential line of thinking has upheld the idea that organizations with structures or forms that 'fit' their environmental context will outperform organizations with forms that do not (in Chapter 2, we referred to this approach as 'contingency theory'). In the strategy literature, this idea has been called the 'challenge of organizational design' or the 'idea of best fit'. Alfred D. Chandler (1962) believed that when an organization had set a goal (i.e. formulated a strategy), this could best be pursued or realized through designing or (re-)structuring the organization. 'Structure follows strategy' thus became a key slogan for organizational theorists (Chandler, 1962; Galbraith, 2002) and strategists.

To take an obvious example, the political and environmental context of the transportation sector does not seem conducive to a strategy of producing fossil-fuel-driven cars in the future. A new strategy of producing hybrid or fully electric vehicles would pose challenges to existing organizational structures, based as they are on the production and distribution of traditional cars. Implementing a strategy to produce low carbon emission cars thus requires some restructuring of car manufacturing companies. Let us take another case: the entertainment industry. The digitalization of the entertainment industry offers two extreme examples of organizations that formulate strategies and establish organizational structures that fit or fail to adapt to the environment. Netflix has made a very successful fit between their strategy and structure, adapting to the digital age, so to say. In contrast, the video rental chain Blockbuster failed to adapt to a new environmental context marked by the spread of digital technologies and went bankrupt in 2010 – attempting to come back later with a new strategy and a new structure focusing on digital products rather than physical ones.

Organizational design

In the field of organizational design, Chandler was among the classical structuralists. Classical structuralists have been described as concerned with the "relatively stable relationships among the positions and groups of positions (units) that comprise the organization. Structural organization theory is concerned with vertical differentiations – hierarchical levels of organizational authority and coordination, and horizontal differentiations between organizational units – for example, between product or service lines, geographical areas, or skills" (Shafritz & Ott, 1978/2005, p. 166).

We recognize this vertical and horizontal thinking in the literature on organizational design. According to Chandler, organizational design has two principal characteristics. First, it seeks to establish lines of authority and second, it creates lines of communication where information and data can flow (Chandler, 1962). Such lines are essential to ensure effective coordination, carry out basic goals, and weave together the total resources of the operation. For instance, a company may have a functional structure, whereby groups that have similar occupational specializations are placed together in units such as Operations, Finance, Human Resources,

STRUCTURE AND INFRASTRUCTURE

Product Research, and so on. The advantages of such a design is that people with similar specialties and skills can collaborate and coordinate easily, thus developing in-depth expertise. A disadvantage of such a design may be that coordination and communication across units may be more difficult, and the different units may develop different understandings of the company's goal.

In practice, the concern with organizational design is discernible in the widely used concept of Business Process Re-Engineering (BPR), which has been developed as a method of optimizing the design of an organization. BPR focuses on analyzing precisely how task execution and resource spending take place, to be able to redesign the organization in the most efficient way. Information technology facilitates such redesign projects because it allows some processes to be rationalized; nevertheless, IT should not be seen as the center of the process. BPR is essentially a conceptual project. As Michael Hammer and James Champy, two proponents of BPR, explain: "Merely throwing computers at an existing business problem does not cause it to be reengineered. In fact, the misuse of technology can block reengineering altogether by reinforcing old ways of thinking and old behavior patterns" (Hammer & Champy, 2003, p. 87).

Organizational designs are often reflected in 'organigrams' or organizational charts. Such charts have a long history in organization studies, as they have been used to depict the units of an organization as well as the relationships between these units (Figure 4.1). The charts can be seen as helpful tools to clarify and interrogate responsibilities within the organization, and as a basis for discussing or developing organizational coordination and communication (Vikkelsø, 2016).

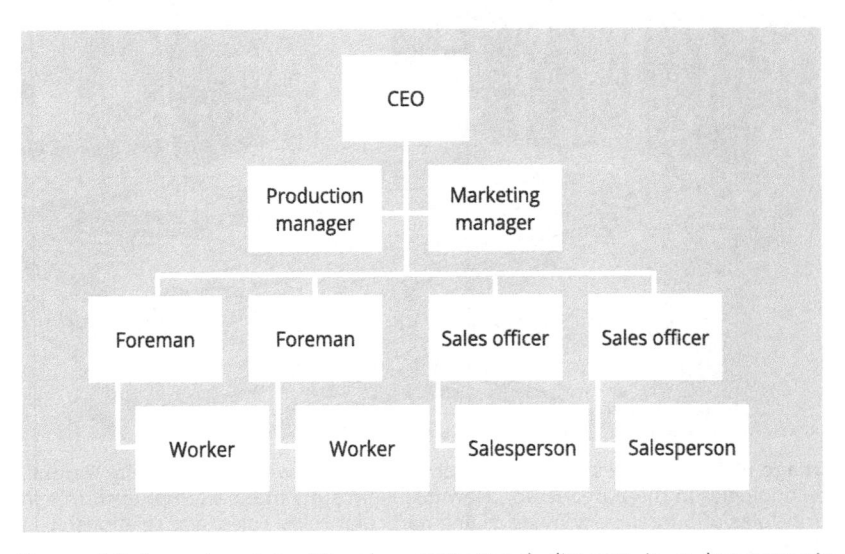

Figure 4.1 Example of traditional organizational diagram in a bureaucratic enterprise, depicting levels and functions.

Bureaucracy

From a different angle, in his work on bureaucracy, Max Weber (1946) offered a view of structure as the primary vehicle for coordination, and as a means of avoiding arbitrariness and inconsistent individual action as drivers of the organization. Weber emphasized the centrality of rules, top-down authority, and specialized roles, as well as documented patterns of information exchange (the files) in running organizations. This way of structuring work was seen as central to modern organizations, both public and private. Their efficiency and legitimacy were dependent on the rationalization tied to bureaucratic structures. Although bureaucracy has been seen by many observers and some social theorists as a dehumanizing way of organizing, and as an inflexible system of procedures that does not allow for special cases, bureaucratic structures have defenders who argue that problems of complexity and uncertainty are not solved by human-centered procedures or charismatic, entrepreneurial management (Du Gay, 2000). In public organizations, the bureaucratic structures and built-in ethos act as a guarantee that public funds and state power will be used in a fair and equitable manner. A bureaucratic organization implies that a court trial is not dependent on a judge's mood or that the allocation of social welfare benefits is not a function of a social worker's personal idiosyncrasies (Image 4.1).

Along these lines of thinking, 'modern structuralists' of the 1960s and 1970s viewed organizations as rational institutions, the purpose of which

Image 4.1 Paper files and their correct storage were traditionally essential technologies in the bureaucracy. New digital systems often attempt to mirror the principles of bureaucracy by inscribing particular roles, rules, and relationships in their architecture.

Source: Sanjeri, iStock.

STRUCTURE AND INFRASTRUCTURE

was to accomplish established objectives. In fact, for Blau and Scott (1962) among others, a purpose or goal is central to the very definition of an organization. An organization is *deliberately* established. It has a specific purpose, as opposed to other types of social groups that can emerge anywhere under other circumstances. For modern structuralists, organizational control and coordination are the tools for pursuing organizational goals, and both control and coordination rely on systems of defined rules and formal authority. In these accounts, hierarchical relations are created through particular structures that ensure coordination and control through the construction of subsystems. These subsystems "are subordinated by an authority relation to the system they belong to" (Simon, 1962, p. 468).

Organizational forms

For the modern structuralists, specialization and division of labor were factors that would ostensibly increase the quality of an organization's output. A central concern was precisely to ensure the quality of the organization's production or functioning. Hence, organizational problems were seen as the result of organizational flaws (Shafritz & Ott, 1978/2005). This is also reflected in Henry Mintzberg's (1979) well-known book, *The Structuring of Organizations*. As a management scholar, Mintzberg was interested in describing how organizations actually structure themselves, and in uncovering how a number of factors can be grouped into ideal type structures, each with their respective strengths and weaknesses. Mintzberg identifies five such ideal typical structures: (1) the simple structure, (2) the machine bureaucracy, (3) the professional bureaucracy, (4) the divisionalized form, and (5) the adhocracy.

The simple structure is often found in young, entrepreneurial companies where an owner or strong leader makes most of the decisions. As long as the company is small, this allows for efficient control and a great deal of flexibility. As companies grow, however, it is often necessary to develop new structures that can facilitate delegation of decision-making power. The second type of structure, machine bureaucracy, has a tight vertical structure where decision-making is centralized and functional lines extend from the top all the way to the bottom. Work routines in a machine bureaucracy are based on a large number of rules and regulations. This kind of structure is found in public organizations such as government agencies and in some large manufacturing companies where procedures and specifications play a significant role in ensuring the uniformity of a product. Type three – what Mintzberg calls the professional bureaucracy – is similarly characterized by a number of rules and procedures, but in contrast to the machine bureaucracy, it tends to employ highly trained professionals who are allowed to make their own, decentralized decisions. This type of structure is typical of organizations such as universities, hospitals, or law firms, where managers cannot make all the decisions.

Type four, the divisionalized organization, has a central headquarters that supports a number of autonomous divisions, each with their own decision-making structures. This kind of organization is typical of large, mature companies, which have several brands, product lines, or which operate in different geographical areas. Finally, the adhocracy has looser structures, since these operate by continuously defining new tasks and assigning individuals or teams to work on these tasks. Ad hoc structures tend to be found in creative industries or in companies whose key objective is innovation. Mintzberg emphasizes that a given organization may have traits from several of these structural forms. Managers, he says, may understand their organizations better if they understand the forces that pull in the direction of different forms, and avoid obvious organizational flaws. Organization scholars have also pointed to organizational forms that are combinations of these structural forms such as the matrix structure, which has elements of both the functional and the divisional forms (Miles & Snow, 1986).

The critique of structures

The idea that there exists a 'best structure' or a 'best fit' for a given type of organization has been challenged both internally in organization theory (e.g. Lawrence & Lorsch, 1967; Woodward, 1965) and by the emergence of new organizational forms: "Terms such as post-industrial, post-bureaucratic, network, virtual, co-evolved, modular and the like, have been minted to try to capture key aspects of such new forms" (McGrath, 2006, p. 582). In her discussion of organizational structure, Rita McGrath points out that the scholarship in this domain has offered a landscape of possible structural alternatives, but no really solid knowledge about 'best fit'. Furthermore, if many structural alternatives are possible, and if it is difficult to create a 'best design' or 'best fit', it may become necessary to reconsider the term structure. Rejecting a static view in favor of a process view, McGrath suggests that organizational 'structures' become less interesting than the idea of 'structuring' as an ongoing managerial activity. This takes us back to the idea of organizational structures as a practical concern; organizational structures are not 'just there'. They need to be reflected and acted on.

Although many scholars have challenged the idea of structures, the types of bureaucratic structures described by Weber have proven to be quite enduring, also in organizations that are being continuously restructured and which have digitized some of their processes. The relatively few organizational studies that treat digitalization of the public sector have shown that even though the public bureaucracy undergoes changes due to digitalization, its basic structure remains in the daily organization of work. In his study of the British Library, Martin Harris (2008) shows that 'virtualization' does not necessarily alter basic organizational structures. He shows that several features of the 'classic' bureaucratic organization remain such as hierarchical management and even a trend toward the centralization of knowledge in the British Library databases. For Harris, it is simply wrong to view digitalization as 'the end of bureaucracy'.

A similar argument is advanced by Catherine Turco in her ethnographic account of 'TechCo', an American social media marketing company that seeks to rethink "the standard playbook for how you're supposed to run a company" (2016, p. 1). As an organization, TechCo is characterized by its radical attempt to eliminate all bureaucratic structures in favor of open conversations on social media (hence, the title of Turco's book: *The Conversational Firm*). For instance, instead of having legal documents governing matters like parental leave and vacations, these are negotiated through an informal and ongoing dialogue between managers and employees. However, as Turco convincingly shows, the absence of formal structures does not necessarily grant employees more influence on decision-making processes. Authority still rests with the TechCo management despite the elimination of formal structures, so the hierarchical element of bureaucracy persists. What *disappears* is the transparency traditionally provided by formal rules and regulations. Similar observations have been made by other scholars concerned with post-bureaucratic modes of organization in the technology sector (e.g. Kunda, 1992).

Organization theory thus seems to leave us with a set of very relevant concepts – authority, hierarchy, and coordination – with which to describe organizations, although we might need to acknowledge the need for concepts that better capture ongoing reorganizations, shifts toward more dynamic lateral relationships, and organizations that now operate as more loosely coupled networks rather than self-contained organizational unities. This is the project suggested by proponents of 'post-bureaucratic organizations' (Heckscher & Donnellon, 1994).

New structures and concepts

As already indicated, organizational scholarship has moved in the direction of trying to capture new organizational forms – 'new', for instance, in relation to Mintzberg's five ideal typical forms. Here, we will discuss two lines of thinking that have become prominent ways of reconsidering or renaming organizational structures: networked organizations and virtual organizations. The networked organization emphasizes the role of horizontal relationships, and that organizational phenomena cannot be contained within the boundaries of an organization. In the virtual organization, we shall see that there is no physical infrastructure to underpin organizational structures, and that the work is geographically distributed.

The networked organization

The networked organization is sometimes understood as a type of organization where the collaboration is defined more by the task than by formal positions in the organization. In a networked organization, individuals, teams, or business entities are flexibly assembled to solve tasks. A networked organization thus consists of "clusters of firms or specialized units coordinated by market mechanisms instead of chains of commands" (Miles & Snow, 1992, p. 53). For example, a core organization may link a number of partners to

DIGITAL ORGANIZING

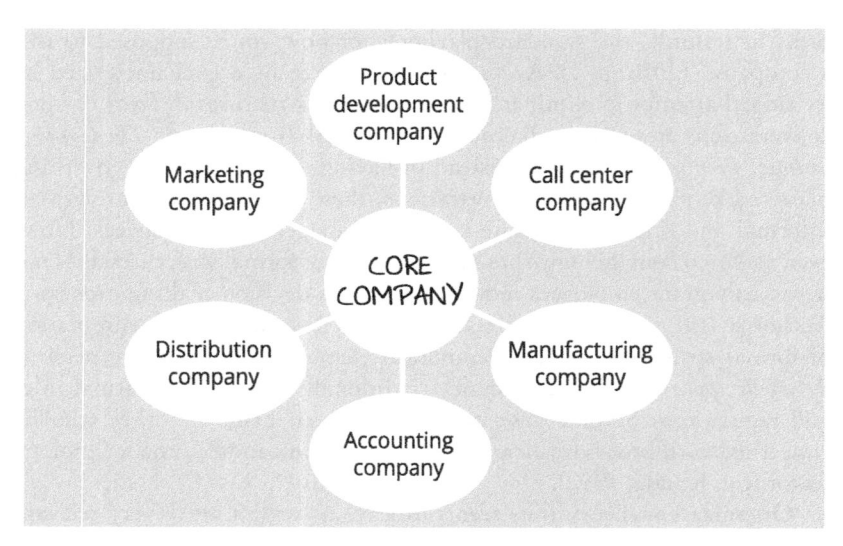

Figure 4.2 Example of network diagram, depicting relationships between business entities.

reach a goal (Figure 4.2). Alternatively, *within* an organization, there may be genuine buying and selling relationships between units. An organic food delivery company, for example, may have close relationships with particular farmers and particular retailers, all of whom are separate businesses, but whose products and services are delivered under a single brand name, relying on the same digital infrastructure and delivery infrastructure.

In Manuel Castells' work, *The Rise of the Network Society* (1996), the idea of the network organization is closely linked to the flow of information channeled by computer networks. Castells describes this as a new type of organization, potentially smaller and more agile, more productive and more competitive than large, vertically structured organizations governed by a bureaucratic logic. For Castells, the power of the network as a loose 'organizational structure' lies in its flexibility, and in the fact that in order to participate, network members must have something to contribute. If it is indeed the case, as Castells argues, that networked organizations are becoming increasingly common, this development will have wide implications for management control and working conditions. If structures are loose and organizational processes dynamic, managers will no longer be able to control flows of information and other important resources (Castells, 1996). Workers will become more individualized because they are not bound together by their place in the organization, but instead connected to projects on an ad hoc basis. Hierarchies become replaced with teams. As Castells writes: "Who are the owners, who the producers, who the managers, and who the servants becomes increasingly blurred in a production system of variable geometry, of teamworking, of networking, outsourcing, and sub-contracting" (Castells, 1996, p. 506).

Common examples of networked organizations are knowledge-based firms (for instance, language and communication services, IT support, personal counseling, or business consultancy firms) or small-scale production or import companies. In the case of knowledge-based firms, these are sometimes physically based in shared office spaces, where office mates can help each other by referring customers or swapping services (Image 4.2). These types of organizations can either expand out of the shared office space and perhaps come to resemble more traditional organizations – as Mintzberg suggested – or they can survive for a long time by just being profitable enough to support the owner and a handful of project-linked helping hands. We also see the rise of 'project organizations', that is, temporary collectives of project stakeholders, project managers, and project teams pursuing a particular goal. The proliferation of such organizations has led some to propose that we live in a 'project society' (Jensen, 2012) and others to problematize the explosion of 'precarious labor'. Networked organizations thus raise important normative questions – but they also raise questions about the relevance of 'organizational structures'.

Castells uses the term 'variable geometry' in a kind of variation on the organization charts we discussed above. His concept suggests that interconnected boxes or circles – and the lines that serve as connectors – cannot adequately describe or help us understand today's organizations. Network organizations are simply too fluid and too connected to actors outside the organization that an image of a bounded organization could be relevant. The border between what is inside and what is outside the organization is too porous, or non-existent.

Image 4.2 Work in shared office spaces around new digital platforms are common organizational activities in what has been termed 'the network society' or 'the project society'.

Source: Metamorworks, iStock.

Event planning: by week	1	2	3	4	5	6	7
Conceptualize Event							
Layout Logistics							
Select Vendors							
Hire Venue							
Hire Caterer							
Hire Event Decorations							
Hire Publicist							
Hire Designer							

Figure 4.3 Example of flow diagram, depicting how tasks are to be solved over time. In project management, such 'Gantt charts' are popular scheduling tools. In this example, the vertical axis shows the tasks relating to event planning, while the horizontal bar chart shows progress on a project schedule. The point of such depictions is that they highlight processes rather than structures.

The popularity of networks – in both academic and public discourse – may partly explain the explosion in flow charts, bubble charts, and other graphic depictions of processes (Vikkelsø, 2016) (Figure 4.3). In a similar vein, some organizations have also adopted new systems for making sense of relations and governance within organizations that retain the dynamic processes involved. Here digital technologies play a supporting role. For instance, 'Holacracy' is marketed as a system that "replaces the traditional management hierarchy with a new peer-to-peer operating system that increases transparency, accountability, and organizational agility" (Holacracy, 2017, n.p.). The system relies on a software program called GlassFrog – "a cloud-based software tool that makes Holacracy transparent and accessible" (GlassFrog, 2017, n.p.). This digital tool traces all the interactions that take place in an organization and displays them online. By conveying all the information of actual relations and activities, it is meant to both combine and exceed an organization's intranet, directory, organizational chart, and other devices.

Rethinking structures in public and private organizations

The use of more networked forms of organizing has not only had an impact in a business setting – as a means to increase competitiveness – but also in the public sector, which is otherwise known for operating according to rigid structures in order to uphold formal legitimacy. Moreover, the network logic has been an organizing principle of many public-private partnership arrangements. It has become increasingly common to formulate and then pursue political goals by linking up political agencies with advocacy groups, private organizations, and even individuals on an ad hoc basis. Such multistakeholder initiatives have become a strategy of various United Nations agencies, for instance, which have otherwise traditionally been slow to reach political goals through conventional diplomatic means.

When various actors come together to solve specific problems, it has been argued that we see a blurring of boundaries between organizations or perhaps even a breakdown of strict organizational boundaries as we know them. When different parties share data, communicate outside of formal settings and structures on various online platforms, and coordinate in new ways, formal structures and boundaries do not appear as solid as they did in more traditional organizational settings. In their discussion of international advocacy networks, for instance, Margaret Keck and Kathryn Sikkink (1998) describe how new organizations emerge around *issues* and manage to build links among actors in civil societies, states, and international organizations, based very much on new possibilities for information exchange. One example could be the so-called UN-Habitat (a program concerned with sustainable urbanization), which gains legitimacy partly through its structure. UN-Habitat's self-description emphasizes that partners range from "governments and local authorities to a wide range of international non-governmental organizations and civil society groups" (UN-Habitat, 2016, n.p.). This tendency toward horizontal collaboration and data sharing can also be seen *within* the public sector, where formerly unconnected departments strive to collaborate on making public service delivery more effective (Margetts, 2009). For instance, if a citizen has a long-term illness and loses his or her job, it would often benefit their case if health authorities and social authorities could cooperate across their respective 'silos'. If different authorities cannot work together horizontally, citizens end up being sent from one office to another and are asked to provide the same information again and again, an experience that gives a negative impression of public service and could threaten the legitimacy of organizations.

If we turn to private businesses, they obviously also participate in such horizontal networks, but maybe more importantly, they are confronted with new organizational realities marked by 'community and sharing'. Private businesses must master the more networked and dialogic communication practices afforded by digital technologies, both among employees and in society at large. One example is the Danish toy company LEGO, which is becoming increasingly engaged in the user communities forming around its products. LEGO offers online platforms for innovation in LEGO products by which users can share building ideas, and they are present on Facebook. LEGO uses these platforms to connect to societal agendas by designing tools for learning in the area of science and technology, and by partnering with various organizations to pursue sustainability projects. We will discuss such initiatives more in Chapter 9 on legitimacy and transparency – the point here is that many businesses spend considerable resources nurturing horizontal relations with a number of actors outside the organization.

The boundaries between inside and outside may not have disappeared completely, but they are certainly being renegotiated.

Such developments have led some business strategy scholars to suggest that managers need to develop new kinds of management and strategy processes, not least in the light of the explosion in new communication infrastructures and online communities. Managers need to adjust their behavior "in reaction to the opportunities offered by the technology" (Gagliardi, 2013, p. 891). Some even talk about Management 2.0, which incorporates a breaking down of lines of authority and the dissolution of the separation between 'levels' in the organization (Koushik et al., 2009). We will discuss these issues in more depth in Chapter 8 on communication and interactivity.

The virtual organization

The concept of virtuality has various meanings, but in this context we will use 'virtual' to talk about a non-physical organization, that is, where interaction among employees and between the organization and its surroundings takes place online. A virtual organization can also comprise an aggregation of organizations that collaborate around a task without being co-located. The virtual organization obviously has some affinities with the networked organization described above. For instance, it relies heavily on digital communication technologies, which allow for collaboration. And it requires other kinds of coordinating mechanisms and control mechanisms than organizations where employees are physically co-located. However, in contrast to the logic of the networked organization, the virtual organization does not necessarily require a rethinking of organizational structures or charts. Where managers, employees, and other stakeholders are not physically co-located, it might still be necessary to carry out work on the basis of a defined structure with defined roles and tasks.

A virtual organization may thus be traditional in some respects. As Charles Handy (1995) points out, there is nothing new in having work activities without a building as a shared home. For instance, networks of salespeople do not necessarily have a common place to operate from. But still, the complete virtualization offered by mobile technologies means that: "Large parts of organizations are now made up of ad hoc mini-organizations, projects collated for a particular time and purpose, drawing their participants from both inside and outside the parent organization. The projects often have no one place to call their own. They exist as activities not as buildings; their only visible sign is an e-mail address" (Handy, 1995, p. 42). Writing this in 1995, Handy was prescient in pointing to some of the issues that arise when employees are not visible to one another or to their managers. For the employee, it might become harder to feel as secure, to feel as socially stimulated, to be able to relate to co-workers, to maintain a sense of purpose, and to improvise. Managers might miss the unscheduled meetings or informal encounters that are conducive to learning and to gaining the ability to control what takes place in the organization in a hands-on

STRUCTURE AND INFRASTRUCTURE

way. Running a virtual organization means that work must be trust-based. Work-units need to be self-contained and able to solve their own problems. Such an organization would be hard to run if it were to mirror the kind of traditional designs outlined by Mintzberg; it is difficult to divide people into functions when they are not together to cooperate. Therefore, Handy talks about the need to 'reengineer work', that is, to create new organizational architectures made up of relatively independent groupings.

The term 'virtual organization' is also intended to capture the coordination that takes place between distributed teams belonging to different companies working on a shared task but who are not physically together (Figure 4.4). In his study of the B-2 'Stealth' bomber construction, Nicholas Argyres (1999) observed how four companies worked to design an aircraft almost entirely in a virtual environment. The firms used a common-access database to manage part designs, and they used an advanced IT system to perform structural analyses. These information systems allowed them to coordinate design and development activities at a surprisingly low cost. Having established a sort of 'technical grammar for communication', along with social conventions for communication, it was possible for the engineers to avoid some inefficiency in governance and decision-making. For Argyres, the lesson is that virtual or disaggregated corporations can be more efficient than hierarchical organizations. He calls them 'vertically disintegrated organizations' and thus speaks to the scholarship on organizational structures and coordination. In the traditional organization literature on structures, a term like 'disintegrated' would have negative connotations.

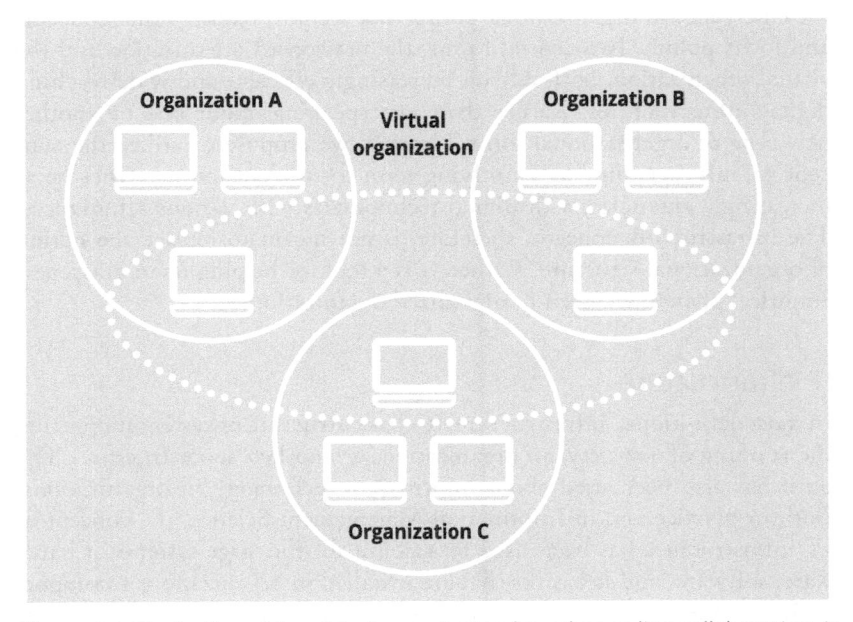

Figure 4.4 Illustration of the virtual organization based on online collaboration. It highlights the importance of digital connections across organizations.

In Argyres' account, however, the replacement of hierarchical structures, with their cumbersome top-down lines of command, is shown to be more productive than problematic.

It seems that even if organizational structures are challenged by virtual organizations it is still relevant to use them as a yardstick against which the virtual organization is measured. How does collaboration work? How are lines of communication ensured? What kind of rules, procedures, tasks, and authorities need to be in place for the work to succeed? A similar interest in organizational structures and concerns can be found in recent scholarship on 'the Blockchain organization' (Scholz & Stein, 2018). Blockchain technology (see also Chapter 1) can be seen as closely connected to the virtual organization in that it offers a new infrastructure for making agreements and sharing knowledge without intermediaries. It is argued that blockchain may lead to the removal of some organizational hierarchical levels, but that competitive advantage will continue to depend on the knowledge and innovative power of employees. As such, blockchain will still require some kind of organization to hold activities together.

Infrastructures

Until now, we have discussed how organization theory has offered a set of concepts that are useful to understand 'organization' as a practical problem of pursuing a shared goal and relating to a given environment. We have also seen that the idea of organizational structure has been challenged by the emergence of organizational forms that seem to require other concepts and focus points. Two central forms, the networked organization and the virtual organization, both rely on increasingly efficient and wide-reaching digital infrastructures. Rather than just speaking about one or another new type of organizational form, however, we propose to utilize the concept of 'infrastructure' as a unifying term for understanding structure as increasingly entangled with digital technologies with various affordances. The infrastructure concept, therefore, is not meant to replace the notion of organizational structure. Rather, it is a tool for helping us to grasp new empirical phenomena seen in digitalized organizations.

IT infrastructure

In most definitions, 'infrastructure' is a basic structure or system supporting the running of a society, an organization, or another 'superstructure'. The term has also been used about information technology in organizations. Both in practice and in Information Management Science, the concept of IT infrastructure has been used to account for the large systems of hardware, software, and networks that are installed in organizations to support their activities. Professionals with an interest in information management are concerned with the systems themselves as well as with the people who use them to pursue organizational goals. A central aim of information managers is to improve organizational performance by ensuring a proper fit between organizational goals, users, and the information technology, which is either developed or purchased and then installed in the organization. Recall the notion of 'fit' between organizational goal and structure in our discussion of organizational design in the section on "Structure as a theme in organization studies" – here we can detect a similar concern, just involving IT in the design of the organization.

Some organizational scholars have suggested that we conceive of IT as a "thread in the changing fabric of organization" (Zammuto et al., 2007, p. 758). For them, it is important to understand organizational form and function in analyses of IT. They argue that IT has supplanted some of hierarchy's role in coordinating and controlling activities. Previously, information management and exchange were carried out within the framework of hierarchies and divisions. With the advent of extensive IT Enterprise Systems, it has been observed that IT "decreased the need to move information through a hierarchy, allowing people to organize around the work itself" (Zammuto et al., 2007, p. 752). IT access freed people on the lower level of the organization to work on more complex and creative tasks, solving problems right away because the necessary information would be readily available. Raymond Zammuto and colleagues believe that this information access is an important development because of the rise of more horizontal work. Like the scholars of network organizations and virtual organizations, they also refer to new organizational forms that are more flexible and less hierarchical, considering how "unpredictable and unanticipated forms of organizing emerge out of the combination of IT and organization features and practices" (Zammuto et al., 2007).

Whereas IT Enterprise Systems were based on the idea that infrastructure is an *asset* to an organization that would invest in large systems, this view has been challenged by a paradigm where infrastructure is seen as a *service* (referred to as IaaS – Infrastructure as a Service). In this paradigm, large and costly internal data centers are partially replaced by cloud computing, where organizations use external servers and applications for their computing needs, especially when they occasionally need more capacity. IaaS creates other concerns for organizations, however. They now need to balance the advantages of capital savings with the risks attached to outsourcing critical day-to-day data processing and sending sensitive data (IBM, 2017) out of the organization – or 'into the cloud'.

Regarding the internal operations of organizations, there are numerous studies into how new IT systems have implications for hierarchical relations and other issues pertaining to organizational structures, the classic statement being Shoshana Zuboff's (1988) *In the Age of the Smart Machine*. Zuboff showed that information technology poses a challenge to hierarchical structures. It is a challenge because the informating function of IT makes information on work processes visible not only to managers, but also to employees, who can now take more control over their work (see also Chapter 10). For instance, when administrative work processes are digitalized, there is less need for systematic control by a manager looking over an employee's shoulder; any case can be made the object of random control and traced back to an employee (Plesner & Justesen, 2018).

A number of empirical studies have demonstrated how IT redistributes work (e.g. Vikkelsø, 2005). To take an example, the introduction of electronic patient records in hospitals has received considerable attention. Studies have shown that when electronic requests and clinical information can suddenly travel across clinical boundaries, this has consequences for both professional work and power relations (Petrakaki et al., 2016). For instance, some nurses have experienced an enlargement of their professional role because they have become responsible for embedding technology in their work, monitoring data, and ensuring proper patient care. At the same time, an established hierarchical relation between doctors and nurses is reinforced, because doctors refuse to engage with the new technology, allowing themselves to consider it non-clinical practice. While the nurses consent to undertaking doctors' digital tasks, this 'assistance' should not be interpreted solely as a sign of submission. It could also be a means by which the nurses maintain and enhance their professional identity. They have now become "patient data custodians and curators" (Petrakaki et al., 2016, p. 220). Recall the discussion of hierarchies in the discussion of formal structures – here, we see that hierarchical relations are not just established formally or by tradition. They have become interwoven with digital infrastructures.

Infrastructure in a broader sense

Using the concept of IT – information technology – points to the centrality of 'information' in this understanding of infrastructure. As used here, 'information' is the translation of objects or practices into a formal representation that can then be shared via information technology. Although information is an extremely important element in organization, using the term in the more technical sense has the drawback of reducing the concept of infrastructure to a purely technical element. In our view, it is beneficial to look beyond this understanding and to use the concept of infrastructure in a more inclusive way. We suggest that to better understand an organization, it is useful to conceive of the infrastructure as containing more informal communication platforms and channels, and as stretching well beyond the organization. Our idea of infrastructure thus includes an organization's 'IT infrastructure', but it is also a broader concept. For instance, it might be relevant to explore how employees use social media, even if social media are not formally part of an organization's IT infrastructure. Social media have a great impact on work for several reasons. To begin with, they are commonly known to distract employees (Stanko & Beckman, 2015). Also, social media are interwoven with semi-professional activities that might have both positive and negative impacts on the organization (Colbert et al., 2016) and their functionalities and user interfaces have been used in the creation of Enterprise Social Media (Leonardi et al., 2013). Finally, the use of social media makes it more difficult to uphold control (over the organization's reputation or external communication) and boundaries (Plesner & Gulbrandsen, 2015; Stanko & Beckman, 2015).

We also need to go a step further in our understanding of infrastructures, drawing inspiration from Science and Technology Studies and from recent internet research. As we touched on in Chapter 3, it is problematic to see 'the technological' as a separate domain from 'the social'. Rather than just thinking of information infrastructures in terms of 'tubes and wires' (Bowker et al., 2010), we should think of infrastructures as complex arrangements of technical, social, organizational, and physical elements. From this perspective, there is no simple relationship between changes in infrastructure and changes in organizational forms, although the two are closely interrelated. Martin Kornberger (2016) has proposed that organization design today – where distributed work and innovation have become increasingly important – relies on the design of 'infrastructures of communication, coordination, and control'. Whereas much organizational design theory is firm-centric, 'organizing in the open' (Kornberger, 2016) combines network thinking with an eye for the infrastructure that might support distributed work.

Influential thinking about infrastructures has suggested that they are powerful organizing mechanisms that structure social reality (Bowker & Star, 1999). Infrastructures are *not* powerful because a design of infrastructures will necessarily result in organizational changes. Rather, they are powerful because they are ubiquitous, almost invisible and boring to study (Star, 1999). Because the webs of infrastructural elements are widespread, consisting of heterogeneous elements, and because they operate below the surface, they have unintended effects on organizational practices.

Reorganizing infrastructures: The case of mental health services

Let us take an example of a reorganization where not only IT infrastructure, but also other infrastructural elements, ended up playing a major role in how an organization functioned and its employees and stakeholders interacted. Mental health services have obviously changed throughout history, but let us assume that we have had a relatively long tradition of health professionals taking care of mentally ill people in specific hospitals or clinics. In a study of the 'virtualization' of mental health treatment practices, organizational scholars followed what happened when patients were no longer treated inside a physical building, but at a distance where communication was supported by digital technologies (Bloomfield & McLean, 2003). This kind of treatment required the development of a particular IT infrastructure, but to understand the organizational changes, we also need to look into how the lack of physical location (and thus co-presence of health professionals and patients) had important organizational implications. In the digitalized organization, health professionals could not just assume that patients were there, but needed to keep track of them through an increased level of documentation in standardized formats. The monitoring of patients, previously supported by physical walls and co-presence, was now upheld through 'needs assessments schedules', committee meetings, and the administration of medicine – all part of the infrastructure of psychiatric healthcare.

Health professionals needed to act as managers of information. Hence, an important new task became that of ensuring sufficient correspondence between the staff's work and their patients' lives. This effort would be bound up with many other issues than their patients' treatment or their virtual connection to healthcare professionals. Digitalization was supposed to compensate for the fact that citizens and public professionals were no longer necessarily co-located. However, the extra time spent on information management actually took away resources from face-to-face professional work, and there were additional issues 'outside of the organization' – in the streets or in private homes – that affected the patients and their care. These are important issues; changes in infrastructure have organizational implications for the delivery of public service, the treatment of mentally ill people, and for the professional identities of health professionals.

Understood in this broad sense, as comprising information/IT infrastructures and other less technical elements, infrastructures become difficult to analyze. They cannot be understood as a 'thing' to uncover but as a configuration of changing relations (Star & Ruhleder, 1996). Therefore, it has been suggested that infrastructures are best observed and understood when some of their elements are changed or break down (Star & Ruhleder, 1996). In this sense, 'infrastructure' becomes quite an extensive addition to the concept of 'structure' that we can use to grasp some important aspects of organizations.

Structure, infrastructure, and sociomateriality

To sum up, the concept of infrastructure can help us understand digital organizing, but it does not make the concept of structure obsolete. It allows us to understand a range of organizing – maybe even structuring – elements that are not captured by viewing organizations as formal structures that can be adequately depicted on an organigram. To illustrate the analytical usefulness of both the structure and infrastructure concepts, we will turn to the story of Occupy Wall Street. Occupy began as an offline protest movement but quickly became an online phenomenon. In 2011, thousands of protesters responded to a call from a Canadian activist magazine to 'occupy' Wall Street in New York City (NYC). They established themselves in a park in Lower Manhattan, setting up tents and 'peaceful barricades' as a sign of protest against Wall Street's influence on politics in the US. The point was to show the world what 'real' democracy looks like. If we look at Occupy with organizational structures in mind we see that the protesters invented a

STRUCTURE AND INFRASTRUCTURE

series of procedures that would ensure a high level of democratic participation and minimize any kind of internal hierarchy. For instance, an elaborate system of hand signals was developed in an effort to allow everyone to participate in decision-making without necessarily having to speak up, and the so-called 'Human Mic' was developed to ensure that everyone was heard despite the lack of actual microphones (Image 4.3).

In mid-November 2011, the protesters were forcefully evicted from the park by the NYC police. Occupy Wall Street did not completely dissolve after the eviction, but became more of an online phenomenon. Instead of protesting in streets and squares, the movement started to coalesce around a Facebook page with almost half a million followers. If we apply a sociomaterial perspective to the infrastructure now supporting the movement, we see that the affordances of Facebook gradually changed the movement. In the beginning, the administrators would mostly author posts that simply informed followers about demonstrations and happenings that took place around NYC and elsewhere. Slowly, however, the character of the posts changed. Instead of merely using Facebook as a tool for mobilization and organization, the page administrators began to post more partisan political messages that did not represent the movement in its entirety. This development troubled some of the Occupy followers, but all they could do was comment on individual posts alongside hundreds of other comments. They were not able to author posts themselves, and they could not oppose or block the administrators' posts in any way. The digital infrastructure simply

Image 4.3 An Occupy Wall Street demonstration. Occupy's human mic was created to amplify messages in the absence of electronic microphones and can be interpreted as an infrastructural element in democratic participation.

Source: Redux, Ritzau/Scanpix.

did not allow for such actions. Hence, the movement suddenly started to grow hierarchical and undemocratic, giving authority back to a few people. In other words, the digital infrastructure of Facebook seriously curbed Occupy Wall Street's ability to extend their democratic experiment into the online sphere (see Husted, 2015; Swann & Husted, 2017).

The story of Occupy Wall Street shows the potential of maintaining a dual focus on structure and infrastructure, both to continuously understand where authority lies, and to understand what technical and material elements do to organizing. In the park, the protesters were extremely concerned with minimizing internal hierarchies and creating horizontal structures. As a new infrastructure began supporting the movement, Facebook was presumed to constitute a democratic and participatory environment in and of itself. Perhaps the myth of Facebook's horizontal, networked character may have encouraged the protesters to downplay the importance of the procedures that had previously made the movement democratic and participatory (consensus, hand signals, Human Mic, etc.). In the online world, it was simply assumed that Facebook could support those procedures and ensure the movement's horizontality. This turned out not to be the case, however, as new practices and coordination mechanisms made the organization much more hierarchical. The case illustrates that no matter how unstable structures may seem, or how many other aspects of the organization we need to take into account in order to understand them, an analytical attention to organizational structures and issues of coordination, control, and communication let us realize that they have a practical reality. Likewise, analyzing infrastructures allows us to interrogate political and organizing powers that otherwise operate below the surface.

? QUESTIONS

1. What is the relationship between organizational structures and digital infrastructural elements in a given organization?
2. How (or why) do bureaucratic structures and ideals continue to prevail in some digitized organizations?
3. How can an infrastructure be analyzed and developed to support the goal of the organization?
4. What is the role of organizational structures in born digital organizations?
5. What kind of effects, intended and unintended, do organizational design choices or infrastructural elements entail?
6. How do coordination, control, and authority play out in organizations where structures are more fluid or informal?

References

Argyres, N. (1999). The impact of information technology on coordination: Evidence from the B-2 'Stealth' bomber. *Organization Science, 10*(2), 162–180.

Barley, S. R. (1986). Technology as an occasion for structuring: Evidence from observations of CT scanners and the social order of radiology departments. *Administrative Science Quarterly, 31*(1), 78–108.

Barnard, C. I. (1938). *The Functions of the Executive.* Cambridge, MA: Harvard University Press.

Blau, P. M. & Scott, W. R. (1962). *Formal organizations: A comparative approach.* San Francisco, CA: Chandler Publishing Company.

Bloomfield, B. & McLean, C. (2003). Beyond the walls of the asylum: Information and organization in the provision of community mental health services. *Information and Organization, 13*(1), 53–84.

Boudreau, M.-C., Serrano, C. & Larson, K. (2014). IT-driven identity work: Creating a group identity in a digital environment. *Information and Organization, 24*(1), 1–24.

Bowker, G. C., Baker, K., Millerand, F. & Ribes, D. (2010). Toward information infrastructure studies: Ways of knowing in a networked environment. In J. Hunsinger, L. Klastrup & M. Allen (Eds.), *International handbook of internet research* (pp. 97–117). Dordrecht: Springer.

Bowker, G. & Star, S. L. (1999). *Sorting things out: Classification and its consequences.* Cambridge, MA: MIT Press.

Castells, M. (1996). *The rise of the network society.* Cambridge, MA: Blackwell Publishers.

Chandler, A. D. (1962). *Strategy and structure: Chapters in the history of the industrial enterprise.* Cambridge, MA: MIT Press.

Colbert, A., Yee, N. & George, G. (2016). The digital workforce and the workplace of the future. *Academy of Management Journal, 59*(3), 731–739.

Du Gay, P. (2000). *In praise of bureaucracy: Weber, organization, ethics.* London: SAGE.

Du Gay, P. & Vikkelsø, S. (2017). *For formal organization: The past in the present and future of organization theory.* Oxford: Oxford University Press.

Gagliardi, D. (2013). Next generation entrepreneur: Innovation strategy through Web 2.0 technologies in SMEs. *Technology Analysis & Strategic Management, 25*(8), 891–904.

Galbraith, J. R. (2002). *Designing organizations: An executive guide to strategy, structure, and process.* San Francisco, CA: Jossey-Bass.

GlassFrog (2017). Glassfrog – Holocracy made easier. Retrieved from https://glassfrog.com/

Gulick, L. (2005). Notes on the theory of organization. In J. M. Shafritz & J. S. Ott (Eds.), *Classics of organization theory* (pp. 84–92). Chicago, IL: The Dorsey Press. (Original work published 1937.)

Hammer, M. & Champy, J. (2003). *Reengineering the corporation: A manifesto for business revolution.* New York: HarperBusiness Essentials.

Handy, C. (1995). Trust and the virtual organization. *Harvard Business Review, 73*(3), 40–50.

Harris, M. (2008). Digital technology and government in transition. The case of the British library. *Human Relations, 61*(5), 741–758.

Heckscher, C. C. & Donnellon, A. (1994). *The post-bureaucratic organization: New perspectives on organizational change*. London: SAGE.

Holacracy (2017). What is Holacracy? Retrieved from http://www.holacracy.org/how-it-works/

Husted, E. (2015). From creation to amplification: Occupy Wall Street's transition into an online populist movement. In J. Uldam & A. Vestergaard (Eds.), *Civic engagement and social media: Political participation beyond the protest* (pp. 153–173). Basingstoke: Palgrave Macmillan.

IBM (2017). Infrastructure as a service. Retrieved from https://www.ibm.com/developerworks/cloud/library/cl-cloudservices1iaas/

Jensen, A. F. (2012). *The project society*. Aarhus: Aarhus University Press.

Keck, M. Sikkink. (1998). *Activists beyond borders: Advocacy networks in international politics*. Ithaca, NY: Cornell University Press.

Kornberger, M. (2016). The visible hand and the crowd: Analyzing organization design in distributed innovation systems. *Strategic Organization*, 15(2), 174–193.

Koushik, S., Birkinshaw, J. & Crainer, S. (2009). Using Web 2.0 to create management 2.0. *Business Strategy Review*, 20(2), 20–23.

Kunda, G. (1992). *Engineering culture: Control and commitment in a high-tech corporation*. Philadelphia, PA: Temple University Press.

Lawrence, P. R. & Lorsch, J. W. (1967). Differentiation and integration in complex organizations. *Administrative Science Quarterly*, 12(1), 1–47.

Leonardi, P. M., Huysman, M. & Steinfield, C. (2013). Enterprise social media: Definition, history, and prospects for the study of social technologies in organizations. *Journal of Computer-Mediated Communication*, 19, 1–19.

Margetts, H. (2009). Public management change and e-government: The emergence of digital-era governance. In A. Chadwick & P. N. Howard (Eds.), *Routledge handbook of internet politics* (pp. 114–127). New York: Routledge.

Mayo, E. (1933). *The human problems of an industrial civilization*. New York: The Macmillan Company.

McGrath, R. G. (2006). Beyond contingency: From structure to structuring in the design of the contemporary organization. In S. Clegg, C. Hardy, T. B. Lawrence & W. R. Nord (Eds.), *The SAGE handbook of organization studies* (pp. 577–597). London: SAGE.

Miles, R. E. & Snow, C. C. (1986). Organizations: New concepts for new forms. *California Management Review*, 28(3), 62–73.

Miles, R. E. & Snow, C. C. (1992). Causes of failure in network organizations. *California Management Review*, 34(4), 53–72.

Mintzberg, H. (1979). *The structuring of organizations: A synthesis of the research*. Englewood Cliffs, NJ: Prentice-Hall.

Petrakaki, D., Klecun, E. & Cornford, T. (2016). Changes in healthcare professional work afforded by technology: The introduction of a national electronic patient record in an English hospital. *Organization*, 23(2), 206–226.

Plesner, U. & Gulbrandsen, I. T. (2015). Strategy and new media: A research agenda. *Strategic Organization*, 13(2), 53–162.

Plesner, U. & Justesen, L. (2018). Merproduktion af målbarhed – Synlighed og nye ledelsesopgaver. *Tidsskrift for Arbejdsliv*, 20(4), 109–115.

Scholz, T. M. & Stein, V. (2018). The architecture of blockchain organization. *ICIS 2018 Proceedings*.

Selznick, P. (1948). Foundations of the theory of organization. *American Sociological Review*, *13*(1), 25–35.

Shafritz, J. M. & Ott, J. S. (2005). *Classics of organization theory*. Chicago, IL: The Dorsey Press. (Original work published 1978.)

Simon, H. A. (1962). The architecture of complexity. *Proceedings of the American Philosophical Society*, *106*(6), 467–482.

Stanko, T. L. & Beckman, C. M. (2015). Watching you watching me: Boundary control and capturing attention in the context of ubiquitous technology use. *Academy of Management Journal*, *58*(3), 712–738.

Star, S. L. (1999). The ethnography of infrastructure. *American Behavioral Scientist*, *43*(3), 377–391.

Star, S. L. & Ruhleder, K. (1996). Steps toward an ecology of infrastructure: Design and access for large information spaces. *Information Systems Research*, *7*(1), 111–134.

Swann, T. & Husted, E. (2017). Undermining anarchy: Facebook's influence on anarchist principles of organization in Occupy Wall Street. *The Information Society*, *33*(4), 192–204.

Taylor, F. W. (1967). *The principles of scientific management*. New York: W. W. Norton. (Original work published 1911.)

Turco, C. (2016). *The conversational firm: Rethinking bureaucracy in the age of social media*. New York: Columbia University Press.

UN-Habitat (2016). History, mandate & role in the UN system. Retrieved from http://unhabitat.org/about-us/history-mandate-role-in-the-un-system/

Vikkelsø, S. (2005). Subtle redistribution of work, attention and risks: Electronic patient records and organisational consequences. *Scandinavian Journal of Information Systems*, *17*(1), 3–30.

Vikkelsø, S. (2016). Technologies of organizational analysis: Charting 'organization' as a practical and epistemic object. *STS Encounters. Research Papers from DASTS*, *8*(3), 1–24.

Weber, M. (1946). Bureaucracy. In H. H. Gerth & C. W. Mills (Eds.), *From Max Weber: Essays in sociology* (pp. 196–266). Oxford: Oxford University Press.

Woodward, J. (1965). *Industrial organisation: Theory and practice*. Oxford: Oxford University Press.

Zammuto, R. F., Griffith, T. L., Majchrzak, A. & Faraj, S. (2007). Information technology and the changing fabric of organization. *Organization Science*, *18*(5), 749–762.

Zuboff, S. (1988). *In the age of the smart machine: The future of work and power*. New York: Basic Books.

5 PRODUCTION AND PRODUSAGE

> In this chapter, we survey the literature on 'production' in organization studies. We identify a profound shift in the way scholars and organization specialists have viewed work and production in capitalist society. Whereas organization scholars originally maintained a sharp distinction between the activities of production and consumption, as well as between the social categories of manager and worker, more recent techno-optimist observers argue that the proliferation of digital technology has blurred these distinctions to the point where wholly new organizational configurations have emerged. According to these scholars, this technology-driven fusion of roles and activities is best captured by portmanteaus like 'prosumption' (production and consumption) and 'produsage' (production and usage). Such concepts clearly draw our attention to important developments in the field of digital organizing. However, they also blind us to situations where these roles and activities continue to remain distinct. When we conflate categories like 'the producer', 'the consumer', 'the worker', and 'the manager', we surrender our ability to meaningfully identify those who actually produce and consume commodities in the digital economy, as well as those profiting most from this process. Hence, in this chapter, we wish to highlight the need to retain some of the original meaning of organized production. Ultimately, this allows us to ask key questions such as, "Who benefits from the digital economy? And at the expense of whom?"

Why production and produsage?

In early 2011, America on Line (AOL – a major online service provider) purchased the independent news and opinion website Huffington Post in a high-profile takeover, turning the vibrant community of voluntary freelance writers into big business. While the then editor-in-chief of Huffington Post, Arianna Huffington, pocketed a staggering $315 million for the sale of her portal, the thousands of freelancers providing content for the site received nothing (Ross, 2013). A few months later, blogger and labor activist Jonathan Tasini filed a class action lawsuit against Huffington Post, describing Arianna Huffington as a "slave owner on a plantation of bloggers" (Gustin, 2011, n.p.). Tasini demanded at least $105 million in compensation on behalf of 9,000 unpaid contributors, pointing to the fact that "without the bloggers there was no Huffington Post and there would be no sale to AOL" (Gustin, 2011, n.p.). Despite the media frenzy, Arianna Huffington appeared untouched by the prospect of a lawsuit, stating that the plaintiffs had no case. They were not under contract, therefore no contract had been broken. The bloggers voluntarily chose to blog – no one forced them to do it. As she explained: "The key point that the lawsuit completely ignores (or perhaps fails to understand) is how new media, new technologies, and the linked economy have changed the game, enabling millions of people to shift their focus from passive observation to active participation – from couch potato to self-expression" (Huffington, 2011, n.p.).

To strengthen her argument, Arianna Huffington and her supporters compared blogging to appearing for free on televised news programs. People do so for two reasons: either "because they are passionate about their ideas or because they have something to promote and want exposure to large and multiple audiences" (Huffington, 2011, n.p.). As an article in *Business Insider* magazine put it, the bloggers did not "deserve a penny": "What Arianna was providing was a place for writers to advertise their wares – a sort of classified for wannabe writers, except unlike the classifieds, she wasn't charging" (MacNicol, 2011, n.p.). In this account, Huffington is cast as a liberator rather than a slave owner. Instead of projecting the image of someone who reaps the fruits of other people's labor, she is seen by many (including herself) as a digital pioneer who provides the masses with a platform for self-expression.

Following the merger with Huffington Post, AOL quickly began laying off in-house writers and editors in order to eliminate overlaps between the two organizations. Around 900 AOL workers, including several well-esteemed journalists, lost their jobs with little warning, effectively reducing the company's workforce by nearly 20 percent (Gustin, 2011). These cutbacks gave Tasini and his plaintiffs more attention in their campaign, but to no avail. Half a year later, the class action suit was rejected by the court, with the judge arguing that "the principles of equity and good conscience do not justify giving the plaintiffs a piece of the purchase price when they never expected to be paid" (Stempel, 2012, n.p.). Since then, AOL has strengthened its focus on the aggregation of content, rendering some of its sites difficult to distinguish from a "content farm" – that is, "a site with shallow, non-original stories written specifically to trigger popular search queries and to game Google algorithms into placing the site on the first page of search results" (Ross, 2013, p. 14). If anything, this move contributed further to the feeling of devaluation experienced by the bloggers at Huffington Post.

The events surrounding the sale of Huffington Post provide a telling example of how many contemporary observers address the themes of work and production within digital organizing. While these themes have traditionally been discussed in a vocabulary that clearly separates workers from managers and distinguishes consumers from producers, more recent techno-optimist literature has argued for an understanding of these roles as increasingly intertwined. In the case of Huffington Post, media pundits ridiculed the plaintiffs by noting that they willfully chose to write blogposts free of charge, only later to accuse Arianna Huffington of profiting from slave labor. As one observer put it: "This all could have been avoided had Jonathan Tasini not acted like the homeless guy on the corner who spits on your windshield, rubs it off 'for free' and then demands money for it" (Masnick, 2011, n.p.). What these observers fail to see, however, is that the homeless person may have little choice but to spit on windshields and rub them for a penny. To be sure, writing blogposts at Huffington Post bears little resemblance to the desperation of homelessness, let alone slavery. However, the techno-optimist narrative of digital technology as an enabler of self-expression prevents us from understanding instances where the digital economy indeed enables actual exploitation (Fuchs, 2014). Let us therefore take a look at classical perspectives on production.

Production as a theme in organization studies

Writings on the organization of production are as old as the field of organization studies itself. In fact, it could be argued that the need for an actual 'theory of organization' emerged as a direct result of the increased scale of production that followed in the wake of industrialization (Gulick, 1937). Frederick W. Taylor, heralded by many as the father of organization theory, was particularly concerned with organizing labor in a way that would reduce the cost of production while increasing the level of productivity in industrial manufacturing. The most effective way to achieve this goal, he argued, was through the complete elimination of what he called 'loafing' or 'soldiering', i.e. the practice of deliberately working at a slow pace (Taylor, 1903/1919). At the turn of the twentieth century, management was generally perceived as a discipline guided by inefficient "rule-of-thumb methods" and "unsystematic observations" rather than proper scientific methods (Taylor, 1911/1967, p. 22). Taylor's ambition was to restore the profession's reputation by providing a more scientific foundation for management that would expose the irrational character of many common management techniques and replace them with financially sounder ones. One of the most commonly used techniques among managers at that time was to increase output expectations whenever productivity increased. In other words, as soon as workers began working harder, managers would lower their piece rate to secure maximum profit.

Taylor saw this as the primary cause of systematic soldiering. Not only did the lowering of wages remove the workers' incentive to work hard, it also created an antagonistic relationship between managers and workers, whereby the interests of the former became incompatible with the interests of the latter. Consequently, he argued that managers should stop punishing workers for being productive and begin rewarding them based on an incentive pay system, with workers being paid individually according to the amount of work they delivered (Taylor, 1895). However, the abolition of collective wages was only one small part of what became known as the 'principles of scientific management' (see also Chapter 2). The most groundbreaking of these principles was to transfer responsibility for planning and coordinating labor processes from workers to management. Previously, workers had operated more autonomously, organizing the work among themselves; the role of management was purely to oversee that the standards for output were met. Under scientific management, managers had to assume full responsibility for *organizing* production, while instructing workers to focus solely on the *execution* of production. As such, the most revolutionary aspect of scientific management was to separate intellectual labor from manual labor, thus removing what was left of 'brain work' from the shop floor (Taylor, 1903/1919).[1]

Since its inception, the theory of scientific management, also known as *Taylorism*, has been heavily criticized for dehumanizing the manual workforce and undermining basic societal values (see Hoxie, 1915). However, contrary to the common perception, the theory was originally conceived by Taylor as a positive-sum method for improving conditions for "all the people" (Taylor, 1911/1967, p. 29). By increasing efficiency in industrial manufacturing, Taylor believed that not only managers, but also workers and society as a whole would benefit – profits would increase, wages would rise, and tax revenues would surge. As he enthusiastically proclaimed: "Maximum prosperity can exist only as the result of maximum productivity of the men and machines of the establishment" (Taylor, 1911/1967, p. 12). This intense focus on efficiency in manufacturing meant that the concepts of production and productivity became an integral part of the study of organization. In fact, as Littler (1990, p. 46) argues, at this point in time, the study of organizations in general could easily be seen as representing little more than a "sociology of productivity". As such, what drove this immense scholarly interest in issues of production was a clear sense of what Valérie Fournier and Chris Grey (2000, p. 17) call 'performative intent' – that is, the commitment to "develop and celebrate knowledge which contributes to the production of maximum output for minimum input". Like Taylor, many researchers worked towards this end by searching for ways to eliminate soldiering and other forms of resistance to managerial authority. Even studies that championed a more 'humanistic' approach to industrial organization such as the well-known Hawthorne studies (Roethlisberger and Dickson, 1939), have been shown to align surprisingly well with the performative intent embedded in the theory of scientific management (Baritz, 1960).

---

1   As Braverman (1974) notes, the distinction between intellectual labor and manual labor is today not adequate, since intellectual labor can also be exploited (think of private universities). The real distinction is between management and workers – those who own the means of production and those who do not.

Labor process theory

From the end of World War II and until the beginning of the 1970s, the industrialized world experienced what economists refer to as the 'Golden Age of Capitalism' (Marglin & Schor, 1992). During that period, growth rates surged to record levels, and unemployment fell to the absolute minimum, propelling the average middleclass family into unprecedented economic prosperity. The primary driver of this economic boom, economists claimed, was a continuous increase in labor productivity. Aided by the introduction of new production technology, companies were suddenly capable of producing much more than previously for much less (e.g. Solow, 1957). In fact, as Michael Perelman (2007, p. 18) notes, U.S. businesses operating in the immediate post-war years were "able to produce as much output as it had a decade before, even though it now used 15 percent less capital and 19 percent less labor". The rising standards of living during the Golden Age provided important legitimacy for the 'performative' study of production processes. Since productivity was seen as *the* driver of the economy, public interest – and, hence, research funds – were generally directed at studies that contributed to improving efficiency in industrial manufacturing (Brown, 1992).

The Golden Age did not last, however. After more than twenty years of economic expansion, growth rates suddenly started to decline, eventually resulting in 'stagflation' (a combination of inflation and high unemployment). The economic downturn led several scholars to question the capacity of capitalism to deliver on its promise to bring prosperity to all. While some thought it necessary to call for a more 'humanistic' version of the market economy (e.g. Harman, 1974), others saw the opportunity to reiterate one of the most fundamental critiques of capitalism as such, namely, that of German political economist and philosopher Karl Marx. Of particular interest to the scholars of the latter persuasion was the part of Marx's comprehensive oeuvre known as the 'labor theory of value' (Marx, 1867/1990). Initially conceived by classical economists like Adam Smith and David Ricardo, the labor theory of value holds that the 'objective' value of any commodity is dependent on the amount of labor required to produce it. Suppose that producing the most recent iPhone takes twice as long as producing the most basic toaster, then – in theory – the iPhone is twice as valuable as the toaster. Obviously, there is a problem with this proposition, since the most recent iPhone is worth much more than the most basic toaster, despite the fact that Apple's massive factory in Zhengzhou is capable of producing half a million phones every day (Barboza, 2016). Marx's contribution to the study of production processes was to account for this discrepancy by inserting a distinction between 'labor value' (the amount of labor required to produce a commodity) and 'exchange value' (the value that this commodity is traded for). The difference between the two, he called 'surplus value', which is merely another word for profit. Marx's re-articulation of the labor theory of value not only made its core proposition more accurate, it also illuminated the fact that some kind of exploitation of workers is required to secure a profit. If the value of a worker's labor corresponds precisely to the value

PRODUCTION AND PRODUSAGE

of the commodity the worker is producing, then there will be no room for profit. Surplus value exists only insofar as the exchange value exceeds the labor value. Given that capital accumulation is the "*differentia specifica* of capitalist production", Marx (1867/1990, p. 769) argued that worker exploitation is the most fundamental dynamic of capitalism (Image 5.1).

The Marxist emphasis on the exploitative nature of production was largely neglected by industrial sociologists and organizational theorists working in the post-war era, presumably because of the high standard of living that capitalism was providing. When the stagflation hit in the early 1970s, many scholars (re)turned to Marxist thinking. One of these was the American steelworker and book editor Harry Braverman, who later came to be seen as the main originator of so-called 'labor process theory' (Thompson & O'Doherty, 2009). What made Braverman's writings compelling to organization theorists was not only that he sought to relate Marxist thought to the study of organized production, but also his attack on the perhaps most canonical figure in classical organization studies: Frederick Winslow Taylor. In his book *Labor and Monopoly Capital: The Degradation of Work in the Twentieth Century*, Braverman (1974, pp. 120–121) advanced the controversial proposition that scientific management represents capitalist

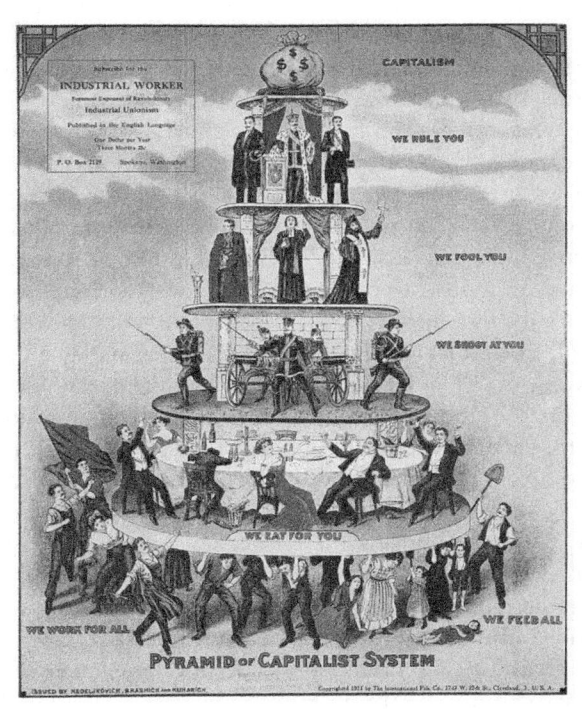

Image 5.1 The pyramid of the capitalist system – a propaganda poster from 1911. The poster criticizes capitalism by illustrating the exploitative relationships inherent to the division of labor in an industrialized society.

Source: Alamy.

thinking *par excellence*, in the sense that it renders "conscious and systematic, the formerly unconscious tendency of capitalist production". What was once a fundamental but non-articulated aspect of industrial organization – the exploitation of workers – was explicitly associated with Taylor's 'optimum' mode of organization.

As we saw have seen, the central principle in scientific management is a complete removal of intellectual labor from the manual workforce, compelling workers to attend solely to the execution of work rather than its conception and organization. This division of labor, Braverman argued, constitutes a fundamental attack on the humanity of workers. Drawing on Marx's theory of alienation, he argued that humans are inherently productive animals. Using tools to satisfy various needs is a fundamental aspect of human nature – one that makes us stand out from most other animals. Moreover, humans continue to produce even in the absence of physical needs (which the never-ending production of iPhones clearly testifies to). However, what truly separates man from animal – what Marx (1844/1978, p. 77) called our 'species-being' – is the ability to conceive the product of our labor *before* actually producing it. As he put it: "A spider conducts operations that resemble those of the weaver, a bee puts to shame many an architect in the construction of her cells. But what distinguishes the worst architect from the best of bees is this, that the architect raises his structure in imagination before erecting it in reality" (Marx, 1867/1990, p. 198).

This means that human work is an inherently creative exercise involving not only the execution of certain operations but also, and sometimes primarily, the conception of these. When workers are instead forced to channel their labor power exclusively toward the execution-part of the labor process, and when the fruits of their toil primarily satisfy the capitalist's needs, then their efforts become indistinguishable from animal work. As Braverman (1974, p. 45) describes it, human work is both "conscious and purposive, while the work of other animals is instinctive". When work becomes a matter of mechanically tightening bolts on an assembly line or shoveling endless amounts of coal (Taylor's favorite example), creativity gives way to instinct, as workers become 'alienated' from the products of their labor as well as their own human nature.

What proponents of scientific management saw as worker 'specialization' (e.g. being an expert in shoveling coal), Braverman characterized as 'deskilling'. By depriving workers of the ability to unite the act of *conception* and the act of *execution* during the production process, workers will inevitably experience a gradual loss of skills or competences, until they finally know how to do only one thing (e.g. shoveling coal). According to Braverman, the root cause of deskilling in the twentieth-century workplace was the introduction of new production technology, which was seen by many as the primary driver of wealth in industrialized societies. Unlike many of his contemporaries, Braverman did not focus on the technical specificities of certain technologies (e.g. size, speed, or complexity), but on their 'social role' in the labor process – that is, the way "a machine supplements man's muscles, mental processes, judgement, and degree of control" (Bright, 1966, quoted in Braverman, 1974, p. 187). What makes more recent technology distinct from, say, a hammer or a typewriter is that it subjects humans to its own rhythm. Workers are thus degraded to a level below technology, in the sense that they must now comply with the machine's mechanic motions and not vice versa (as with a hammer).

Following Braverman's penetrating critique of the capitalist mode of production, labor process theory was further developed by a series of writers who took a less dogmatic stance regarding the oppressive nature of organized production. For instance, Michael Burawoy (1979), based on his own experiences as a machine operator, observed that workers often played an active role in their own subordination by willfully participating in the game of 'making out' (i.e. maximizing incentive pay). In so doing, they not only earn more money by working faster, they also turn tedious work processes into sports-like competitions. This led Burawoy to argue that the exploitation of workers under capitalism greatly depends on their own consent, which was an argument that served to weaken the idea of an antagonistic relationship between the irreconcilable interests of employer and employee (see also Cressey & MacInnes, 1980; Knights & Willmott, 1989; Willis, 1977). In contrast to Braverman's argument, Burawoy found that technology did not necessarily reinforce worker dehumanization by way of deskilling. Rather, when workers became responsible for the functioning of machines, it often created a sense of autonomy, making them feel in control of their own labor. This juxtaposing of technology and autonomy is worth keeping in mind when considering the literature on digital organizing. As we shall see, the strong focus on the antagonistic relationship between worker and manager that characterized many of the early studies of organized production has gradually given way in the digital era to a focus on the collaborative nature of work – as if digitalization has eliminated class conflict.

The rise of the prosumer

Initially, the study of organized production was closely connected to the industrial age and the problems associated with assembling many 'producers' in the same location. Having legions of people working alongside each other in large factories sparked an unprecedented need for management and organization, which eventually led to the artificial separation of brain work and manual labor, as we saw above. However, when the industrial society became 'post-industrial' two-thirds into the twentieth century (Bell, 1974), and when scholars suddenly started talking about the world as a 'global village' (McLuhan, 1964) created by the technologies of the 'information age' (Dizard, 1981), the strong focus on the labor process as a source of antagonism between workers and managers slowly faded. One thinker who had a significant impact on this agenda was the futurist and techno-optimist, Alvin Toffler. In his bestselling book, *The Third Wave*, Toffler vividly describes the pending "death of industrialism and the rise of a new civilization", which will presumably force us to reconsider "all our old assumptions", as these no longer "fit the facts". As he put it: "The world that is fast emerging from the clash of new values and technologies ... demands wholly new ideas and analogies, classifications and concepts" (Toffler, 1980, p. 16).

One of the most groundbreaking concepts to emerge from Toffler's account of the future civilization was the notion of the 'prosumer' (combining the words producer and consumer). In preindustrial society, Toffler argued, the rigid division between producers and consumers was almost non-existent, meaning that the level of commodity exchange was much lower than it is today. Farmers, blacksmiths, carpenters, and other producers almost exclusively manufactured goods for private use, for barter, or for the aristocracy. As such, the consumer did not exist in any meaningful way, because the capitalist marketplace did not exist. The technological revolution of the late eighteenth century and the introduction of a formal monetary system (the gold standard) violently reconfigured this condition by "driving a wedge into society", separating the function of the producer from that of the consumer (Toffler, 1980, p. 277). Instead of manufacturing products for their personal or household use, people now began manufacturing products for exchange. According to Toffler, this signaled the death of the preindustrial prosumer.

The prosumer remained dead for approximately 300 years, until new technology paved the way for yet another revolution, referred to by Toffler as the 'Third Wave of change' (the agricultural revolution being the first, industrialism the second). In this new age, producer and consumer once again merge into a single individual, as people start producing for themselves rather than for others. Writing in the late 1970s, Toffler saw signs of this grand transformation everywhere: self-service gas stations, electronic banking, 24-hour customer service, do-it-yourself trends, self-help movements, computer-aided manufacturing, and much more. The common theme across the multitude of cases described in Toffler's book is that consumers now become active participants in the production process. In contrast to Henry Ford's (1922) mantra that customers "can have a car painted any color that he wants so long as it is black", consumers gradually seemed to become co-producers themselves. As Toffler (1980, p. 285) noted in relation to the prospect of tech-savvy customers designing their own products through computer-aided manufacturing systems: "In the end, the consumer, not merely providing the specs but punching the button that sets the entire production in action, will become as much part of the production process as the denim-clad assembly-line worker was in the world now dying."

Toffler went so far as to conclude that the rise of the prosumer would eliminate one of the most defining features of the industrial age – the process of marketization. Once consumers start utilizing new technology to manufacture products for themselves, the need for a money-based market to coordinate exchange will diminish, eventually leading to the twilight of consumption. According to Toffler, a by-product of this massive societal change is that dominant economic models will become redundant because of their inability to explain the forces at work. Since all these models are predicated on a split between production and consumption, as well as a split between capital and labor, the fusion of these activities demands a completely new vocabulary capable of addressing the theme of organized production in post-industrial society – that is, a vocabulary for "a civilization able to move on to a new agenda" (Toffler, 1980, p. 298).

Production in the informational economy

A decade and a half after Toffler's call for a new vocabulary, the Spanish sociologist Manuel Castells published a pioneering trilogy, called *The Information Age*, on the societal changes brought about by the transition from an industrial economy to an informational economy. Castells seriously challenged Toffler's claim that we are witnessing a fundamental alteration of capitalism. Instead, he argued that the informational economy largely subscribes to the same logic as the industrial economy and that it sometimes even exacerbates it (see also Kumar, 1995). However, what makes the informational economy different, according to Castells, is that knowledge moves to the center of the production process. In the industrial economy, knowledge was a means to achieve a more efficient production process. In the informational economy, knowledge itself becomes "the product of the production process" (Castells, 1996, p. 67).

One of the most influential theorists to back this claim was the Japanese organization scholar, Ikujiro Nonaka, whose work has since become known for establishing the centrality of knowledge in the informational economy (see also Chapter 7). Based on observations of successful Japanese corporations such as Honda and Canon, Nonaka (1991, p. 96) argued that "in an economy where the only certainty is uncertainty", knowledge creation is the most effective way of maintaining a competitive advantage. While some companies may become successful based on their ability to produce large quantities of products in an efficient manner, their inability to adapt to changes in the environment (new consumer demands, new production techniques, etc.) will soon turn success into failure. Nonaka's main thesis, therefore, was that companies should be viewed as ever-changing organisms rather than as static machines, and that workers should be treated as the locus of knowledge creation rather than as simple cogs in the machine.

Underlying this thesis was the concept of 'tacit knowledge', which represents the type of knowledge that resides within an individual, but cannot be readily expressed (an oft-used example is bicycle riding: it is fairly easy to do, but rather difficult to explain). The way to render this form of implicit knowledge explicit, Nonaka argued, was to include ordinary workers in the conception-part of the labor process from which they had previously been excluded. For instance, he advised companies to structure their organization into a number of "self-organizing teams", in which employees should be given more "freedom and autonomy to set their own goals", as this would supply them with the incentive to share personal insights with the team as well as the rest of the organization (Nonaka, 1991, p. 104). Evidently, this new type of 'Japanese' management constituted a violent attack on the rigid division of labor between brain work and manual work that characterized most of the early studies of organized production. In the words of Castells (1996, p. 160), the popularity of the Japanese model created a "profound transformation of the management-labor relationships".

Although digital technology did not play a direct role in Nonaka's theorization of organizational knowledge creation, Castells claimed that he and Nonaka had privately discussed how "on-line communication and computerized storage capacity" had become "powerful tools in developing the complexity of organizational links between tacit and explicit knowledge" (Castells, 1996, p. 160). Castells then used their discussion as a pretext for extending Nonaka's work by introducing what he called the 'network enterprise' (see also Chapter 4, which discusses different conceptions of networked organizations). Castells' central idea about the network enterprise is that it substitutes traditional hierarchical structures for more horizontal and networked structures, in an effort to "adapt to the conditions of unpredictability ushered in by rapid economic and technological change" (Castells, 1996, p. 164). In the network enterprise, workers are freer and better informed than previously; they work more independently and use digital technology to coordinate their activities among themselves; but most importantly, they are invited by management to participate in conceiving the production process. In many ways, the network enterprise is the negative image of Taylor's industrial establishment.

Individualized production

With the rise of network enterprises, the antagonistic relationship between capital and labor that Marx (1867/1990) emphasized as crucial to understanding the exploitation of workers under capitalism, an antagonism that Braverman (1974) identified as the cornerstone of scientific management, was blurred to the point of near-conversion. If workers suddenly become 'autonomous' and 'self-organizing' in the network enterprise, the previously palpable distinction between employer and employee vanishes. Castells characterized this development as a fragmentation and individualization of the workforce. If the industrial economy *socialized* work by gathering many producers in the same location, the informational economy *individualized* work by dispersing producers across time and space. This lack of physical connection and social cohesion in the workforce gave workers the freedom to structure their work more independently. However, it also made it difficult for them to protest poor working conditions and low wages, as the common ground from which to protest had disappeared.

For Castells, these developments are tightly linked to technology. Without digital information and communication technologies, workers would not be able to interact and collaborate independent of time and space constraints; without advanced manufacturing technology, corporations would not be able to adapt to constant changes in the business environment; and without powerful data-processing technology, managers would not be able to detect these changes in the first place. Hence, it is not surprising that the rise of the network enterprise coincided with, and was even dependent on, "the massive diffusion of information technology" (Castells, 1996, p. 242).

From prosumption to produsage

At the turn of the millennium, 'the consumer' was officially pronounced dead by a young American technology writer named Clay Shirky. What had hitherto seemed like a heated dream in Toffler's futurist mind had apparently come true 20 years after the publication of *The Third Wave*. In a bold text entitled "RIP the Consumer, 1900–1999", Shirky advanced the surprising claim that the consumer had become "the internet's latest casualty" (Shirky, 1999, n.p.). The reason for this largely unforeseen event was that the interactive nature of online communication had undermined the almighty role of mass media. In the industrial age, Shirky claimed, the consumer was "nothing more than a giant maw at the end of mass media's long conveyer belt, the all-absorbing Yin to mass media's all-producing Yang". Whatever corporations produced and advertised on TV or radio, ordinary people would passively consume.

With the rise of the internet, however, the passive consumer turned into an active media outlet. Instead of uncritically purchasing whatever products were advertised to them, people suddenly began talking back to corporations, writing complaints and posting demands in the digital public sphere. As such, corporations were no longer protected from the wrath of unsatisfied customers, and with their reputations threatened, all they could do was comply. At least, that is how Shirky interpreted the diffusion of digital technologies such as online blogs at the end of the twentieth century. He thus arrives at the sweeping conclusion that, rather than being a herd of passive consumers, "we are all producers now" (Shirky, 1999, n.p.). Shirky's dramatic account of the causes leading to the death of the consumer gained little traction within the academic community in the immediate years following its publication in 1999. By 2006, however, following the launch of participatory platforms like YouTube, Facebook, MySpace, and Wikipedia, the idea that the public had become active producers seemed to register more widely. One of the most memorable signs of this was *Time Magazine*'s decision to declare 'You' as Person of the Year, publishing a mirror on the cover page and stating that "you control the media now, and the world will never be the same" (Grossman, 2006, n.p.).

That same year, two influential business consultants tried to capture this transformation under the heading 'wikinomics', as a way of describing how the economy had come to resemble the collaborative infrastructure of wikis (Tapscott & Williams, 2006). Though subscribing to the same assumptions about the game-changing role of the internet as Shirky, Don Tapscott and Anthony Williams pursued a more radical argument. While most observers considered ordinary people 'producers' to the extent that they participate in the creation of *online content* (YouTube videos, blog posts, Facebook updates), Tapscott and Williams argued that such activities "are really just the tip of the iceberg", and that an entirely "new mode of production is in the making" (Tapscott & Williams, 2006, p. ix). In this new mode of production, customers are not just actively engaged in "talking back" to corporations by way of complaints and comments; they now insert themselves in the front end of the value chain, in order to "participate in the design, creation, and production of the product" (Tapscott & Williams, 2006, p. 125). As an example of this emerging trend, Tapscott and Williams used the massive multiplayer online game *Second Life*, which is largely created by its customers. Second Life players use personalized avatars to navigate a virtual world, in which they can manufacture virtual objects and trade them through an internal marketplace. As such, they do not just modify and customize already existing products, as Shirky envisioned, but create them more or less from scratch.

Despite the ambition to describe an entirely new paradigm of organized production, Tapscott and Williams retained Toffler's old concept of 'prosumption' to describe how, in the world of wikinomics, the "gap between producers and consumers is blurring" (Tapscott & Williams, 2006, p. 125). According to the German media scholar Axel Bruns, this was a mistake, as it preserved the idea that it remained possible to demarcate the activities of production and consumption in any meaningful way – something that he forcefully rejected. In fact, as Bruns (2008, p. 12) notes in his book *Blogs, Wikipedia, Second Life and Beyond*, the notion of prosumption merely describes "the perfection of the feedback loop from consumer to producer" rather than an actual blending of the two roles. To remedy this misconception, Bruns coined yet another term – 'produsage' – as a portmanteau of production and *usage*. He thus definitively kills off with the concept of consumption. People no longer simply 'consume' things online, they 'use' them (see also Chapter 8 on interactivity), and in that process, Bruns maintains, lies a creative potential. As he puts it, "consumers themselves are, of course, no longer just that, but active users and participants in the creation as well as the usage of media and culture'" (Bruns, 2008, p. 16) (Figures 5.1, 5.2, and 5.3).

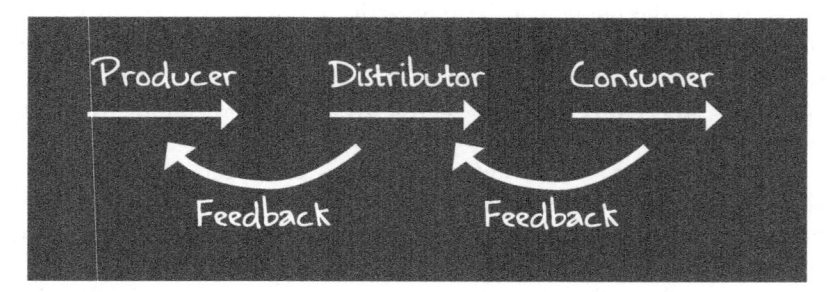

Figure 5.1 The traditional (industrial) production–consumption process with limited feedback loop from consumer to producer.

PRODUCTION AND PRODUSAGE

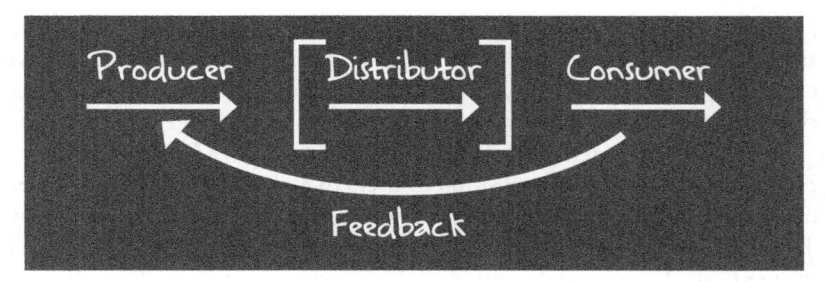

Figure 5.2 The prosumption process, where the feedback loop runs directly from consumer to producer.

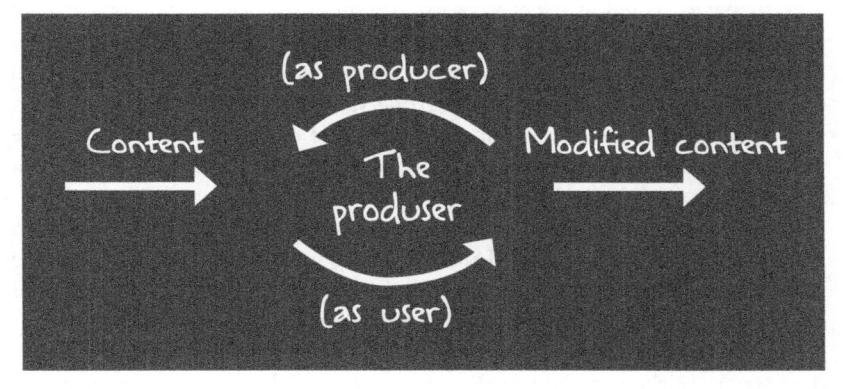

Figure 5.3 The produsage process, where producer and consumer are merged into the 'produser'.

Produsage in practice: The cases of Wikipedia and 3D printing

The example that Bruns seems to prefer is Wikipedia. What makes Wikipedia stand out from many other online encyclopedias is that its editability is not restricted to certain areas or to particular user segments (though popular pages are 'semi-protected' due to the risk of vandalism). On Wikipedia, 'anyone can edit', as the slogan goes. This initially caused a great deal of controversy, with many observers arguing that the platform's open-source approach to knowledge production undermined the encyclopedia industry's claim to truth and objectivity. However, after a study in the highly respected scientific journal *Nature* claimed that entries in Wikipedia were almost as accurate as entries in the expert-based *Encyclopedia Britannica*, the open-source approach suddenly did not seem as reckless as before. Both encyclopedias contain inaccuracies, the study found, but due to its open participation, Wikipedia corrects mistakes much faster (Giles, 2005). This leads us to the first of four key characteristics of produsage, highlighted by Bruns. According to him, production processes that meet these four requirements are not production processes in the traditional (industrial) sense, but proper produsage processes. The first such requirement is called 'open participation, communal evaluation', the point being that in

119

produsage processes, anyone can participate in the creation and examination of content. In the case of Wikipedia, anyone can create and edit an entry. Furthermore, by clicking the 'talk' button, participants can discuss the content of a page and evaluate its quality.

The second characteristic is called 'fluid heterarchy, ad hoc meritocracy'. Here, the point is that the position of users in a produsage process is determined by their skills (merit) rather than by some kind of fixed hierarchy. The fluid nature of this system means that users who are highly influential in one part of the process may be less influential in others, depending on their actual skills and ideas. On Wikipedia, some users are more active than others, and these users often acquire a position of high status in relation to specific pages, as fellow users recognize their contributions. Though official administrators exist on Wikipedia, Bruns (2008, p. 109) does not consider them a threat to the merit-based structure of the overall project. The third characteristic is called 'unfinished artifacts, continuing process', which is a way of describing the always incomplete nature of products in a produsage process. Following from the above, entries on Wikipedia are never finalized, but always a work in progress.

The fourth and final characteristic of produsage is called 'common property, individual rewards'. In order to grow, Bruns claimed, produsage processes need to generate some type of individual rewards. Users do not labor for purely altruistic reasons. At the same time, however, the content created through such processes should rely on shared ownership. Clearly, the texts published on Wikipedia are not owned by any private individuals, but by the collective (by way of its Creative Commons license). What motivates people to produce text for Wikipedia, Bruns argues, is a feeling of contributing to the common good as well as a genuine desire for knowledge creation within particular fields.

Although Bruns illustrates his points mostly by reference to internet communities, the concept of produsage does not refer only to intangible 'content'. It could equally apply to more physical products. Given the right technology, almost anything can be 'prodused', because most things have an informational aspect. As he notes: "In time … it will also become clear that new modes of informational production and produsage have the ability to affect even strongly physically based industries" (Bruns, 2008, p. 13). Or, as former *Wired* editor Chris Anderson (2010, n.p.) puts it: "Peer production, open source, crowdsourcing, user-generated content – all these digital trends have begun to play out in the world of atoms, too. The Web was just the proof of concept. Now the revolution hits the real world." The most obvious example is 3D printing. By purchasing one of the many commercially available 3D printers, 'ordinary people' are now able to circumvent traditional flows of production, thus applying the principles of produsage to physical manufacturing as well (Hamalainen & Karjalainen, 2017). It might be that the actual 'printouts' do not meet the requirements for produsage, as outlined by Bruns (for one, they are rather finished once

PRODUCTION AND PRODUSAGE

printed), but the computer-aided design files used for printing could easily be created in this way.

Such changed dynamics in production are visible on platforms like Shapeways, a 3D-printing marketplace and service. Shapeways allows anybody to design a product, have it 3D-printed on-demand, and sell products to one another. Shapeways does not offer design services. It provides the infrastructure for design, manages the actual 3D production and display of products on its website, and receives payment by the client for materials, printing, marketing, and distribution of the sold objects. Shapeways thus allows people to set up their own little stores online and lets them decide the price of the design (Shapeways, 2017). No central management decides on a product portfolio or on the amount of resources to be spent on design. This is the user's own decision. Shapeways is thus a platform for individual entrepreneurs rather than a production and marketing firm selling a product (Image 5.2).

In that sense, the concept of produsage reaches beyond Second Life and Wikipedia and well into the material world, heralding an era of more autonomous and voluntary production processes where the hierarchies of the industrial economy finally give way to the networks of the informational economy. Produsage is thus a useful way of describing a novel mode of production that is not only user-centered (like Toffler's notion of prosumption) but also user-led. In produsage processes no one tells users what to *consume* through mass mediated advertisements, just as no one tells them

Image 5.2 Work with a 3D printer. Maker spaces are often equipped with tools such as 3D printers. A range of digital technologies allows anyone to become a 'produser' of rather advanced objects or products.

Source: Andresr, iStock.

121

what to *produce* at work. As such, the concept of produsage merges not only the social activities of production and consumption, but – more or less intentionally – also the previously distinct social categories of worker and manager. Workers and managers are now one single identity: the *produser*.

The blind spots of produsage

It is tempting to believe that we live in a world devoid of passive consumption and alienating work, or that we – if nothing else – are on the verge of realizing such a world. For many of the techno-optimists cited in this chapter, the rise of new digital technology seems to provide ample conditions for a much more harmonious existence. For Toffler (1980, p. 394), the dawn of a 'Third Wave society' signals the end of "repetitive, specialized, and time-pressured" labor and the emergence of a new type of work where "people divide their lives between working part time in big, independent companies or organizations and working part time for self and family in small autonomous prosuming units" (Toffler, 1980, p. 298). Similarly, according to Tapscott and Williams, a "truly self-organized and distributed way of working" constitutes "an imminent reality that few workplaces today are prepared for" (Tapscott & Williams, 2006, p. 267), but one that "many employees will welcome … as they search for flexibility, identity, ownership, authenticity, and continuous learning" (Tapscott & Williams, 2006, p. 265). Finally, for Bruns (2008, p. 396), the proliferation of produsage has the potential to create a "new type of civilization that is not exclusively geared towards the profit motive" and forced to "reduce its reliance on hierarchical, top-down forms of social organization". This shift not only fosters more creative and democratic labor processes for the benefit of individual workers, it also constitutes one of "the most important contributions to global development and international welfare" (Bruns, 2008, p. 405).

But maybe these conclusions are too hasty. Concepts like 'prosumption', 'wikinomics', and 'produsage' no doubt allude us to emerging trends that might be difficult to understand with traditional theories of organized production. After all, consumers certainly exercise more influence over production processes than previously, and fewer people are indeed employed in the manufacturing sector than ever before. However, like all concepts, these revolution-promoting concepts have blind spots that prevent us from seeing certain characteristics that may prove crucial to our understanding of digital organizing. Hence, it is not necessarily true that a term like 'production' is hopelessly outdated and "no longer accurate", as Bruns (2008, p. 1–2) suggests, and that we can see things more clearly if we could free ourselves of "the baggage of 'common sense' assumptions and understandings about industrial processes of content production". We might see the world differently, but doing so comes at a cost. In the following section, we will highlight three potential blind spots created by the notion of produsage.

Are we all producers now?

One of the main critiques leveled against the notion of produsage begins with a questioning of the proposition that the passive consumer is dead and that 'we are all producers now', as proponents of the techno-optimist stance proclaim (e.g. Bruns, 2008; Shirky, 1999; Tapscott & Williams, 2006). This proposition rests on the flawed assumption that, just because anyone *can* create content, they automatically *will*. This is precisely the type of technological determinism that we need to question. A number of scholars have already done so. For instance, José van Dijck (2009) argues that user behavior is infinitely more complex than the binary distinction implied by a contrast between passive consumption and active produsage, making this distinction a historical fallacy. Furthermore, the vast majority of supposedly active users are, in fact, relatively inactive in their online behavior. To bolster her point, van Dijck (2009, p. 44) draws on a comprehensive survey conducted by an American research team, which found that:

> Of all online users of UGC [user-generated content] sites, 13 percent are 'active creators' – people actually producing and uploading content such as webblogs, videos or photos. Just under 19 percent qualify as 'critics', which means they produce ratings or evaluations. Furthermore, 15 percent of all users are 'collectors', a term referring to those who save URLs on a social bookmarking service ... another 19 percent count as 'joiners' – people who join social networking sites such as MySpace or Facebook, without necessarily contributing content. The majority of users consists of 'passive spectators' (33%) and 'inactives' (52%).

With 85 percent of all users being either passive spectators or inactives, the techno-optimist narrative of widespread produsage is exaggerated, if not simply fallacious (see Bird, 2011; Comor, 2011; Van Dijck & Nieborg, 2009). A more accurate description might be, as suggested by George Ritzer and Nathan Jurgensen (2010), that the co-existence of traditional modes of production, passive consumption, and active produsage is what truly characterizes our time, rather than an actual predominance of any of them. This critique has been extended by David Brake (2014) who points to the existence of a 'digital divide' separating a privileged minority of content creators from the vast majority of less-privileged inactive users, both globally and within their own national context. In this sense, the archetypical produser should be seen as an elite user who is young, online-skilled, educated, well heeled, and living in the Western world (see also Siapera, 2017).

User commodification

A second critique of produsage is that the term makes us blind to what Dallas Smythe more than 40 years ago called 'the audience commodity'.

In a path-breaking article, Smythe (1977) argued that Western Marxism had turned a blind eye to the way in which TV and radio audiences are sold as commodities to advertisers. During the daytime, he argued, workers are forced to sell their labor power to capitalists in order to make a living. When they return home from work, however, they are still not finished 'working'. By watching TV commercials, listening to radio jingles, or reading newspaper ads, workers continue to labor in the interest of capital throughout the rest of their waking hours. As Smythe (1977, p. 3) puts it: "All non-sleeping time of most of the population is work time." The difference between this type of audience work and 'proper' work is that the former is unknowing and unpaid, making it a capitalist dream scenario. Today, the interactive and personalized nature of web 2.0 platforms has provided a fertile ground for corporations like Facebook and YouTube to collect, package, and sell the attention of its users to corporate advertisers (McGuigan & Manzarolle, 2014). By searching for information, uploading content, and registering personal data, users are constantly providing data that can be traded (see also Chapter 11). Today, the object of commodification is no longer the passive audience, but the active user.

Christian Fuchs (2014) refers to this phenomenon as 'economic surveillance', meaning that corporations systematically monitor the interests and behaviors of internet users in order to target advertisements in a much more individualized manner. If you have ever wondered why advertisements on your Facebook page always mirror your most recent Google searches, or why Amazon always seems to know exactly what type of books you prefer, this is why. As Fuchs (2014, p. 101) notes: "Whereas audience commodification in newspapers and traditional broadcasting was always based on statistical assessments of audience rates and characteristics (Bolin 2011), internet surveillance gives social media corporations an exact picture of the interests and activities of users." This represents a sobering antidote to the assertion that the birth of the produser would usher in a "new type of civilization that is not exclusively geared towards the profit motive" (Bauwens, 2005, quoted in Bruns, 2008, p. 396), which it clearly does not. Moreover, it alerts us to the fact that the content uploaded on various platforms might not be the produsers' most valuable asset, but instead the metadata (personal information, cultural preferences, buying habits, etc.) they unknowingly submit to corporate interests (van Dijck, 2009).

Thus, the concept of produsage (as well as 'prosumption', 'wikinomics', and many more) masks processes of exploitation, as they unfold in the digital realm. We have seen that capitalism is fully dependent on worker exploitation to survive, since this is the most effective way of generating surplus value. In other words, someone must work in an un(der)paid manner for capitalism to persist. Hence, whenever confronted with neologisms like produsage that claim to install harmony between labor and capital, we must always ask ourselves: *Who's benefiting in the digital economy? And at the expense of whom?* As we have seen, one answer to this question is the billions of internet users who, often unknowingly, labor in the interest of capital by watching advertisements, creating content, and providing metadata (see also Zuboff, 2019), but also the millions of sweatshop workers producing the material infrastructure for the digital economy. We will address this issue in the following.

The cyber proletariat

The third and final critique of produsage has to do with its somewhat myopic focus on the intangible side of production. In discussions of user agency and audience commodification, we risk losing sight of the more material dimension of the digital economy. Platforms like Wikipedia or virtual worlds like Second Life need hardware, produced in traditional factories. Indeed, what would be the value of any web 2.0 application without those millions of phones and tablets produced every single day? All the devices that we use daily are made from minerals and metals extracted from the earth. For instance, a typical smartphone contains small quantities of precious materials like gold, silver, platinum, and palladium, but also larger quantities of less valuable metals like aluminum and copper, as well as a range of commonplace but hard-to-extract earth elements, including yttrium, lanthanum, terbium, neodymium, gadolinium, and praseodymium (Nogrady, 2016). The earth's supply of such materials is far from inexhaustible, and in that sense, one could argue that the earth itself is being exploited by our insatiable demand for novel technology.

But that is not the full story. Keeping the environmental cost of digital technology in mind, we must not forget that someone has to extract these precious materials from deep below the earth's surface. As multiple researchers, advocacy groups, and government agencies have recognized, the human cost of smartphone production is incredibly high. For instance, a YouTube video uploaded by the European Parliament (2017) notes that many extraction sites (often located in rural Africa) are controlled by armed groups and that many of the workers are children, making mineral mining a "major cause of insecurity, poverty, and human rights abuses". Not only is it dangerous to enter the ramshackle mines, the amounts of mercury and cyanide used particularly in gold-extraction also present a massive health risk to workers. In addition, people are often forced to work the mines or choose to do so only because they have no other options. Fuchs (2014) characterizes this as slave labor, reminiscent of the type of labor underpinning feudal society.

Horrifying as it may seem, this is once again not the full story, since someone has to turn those minerals, extracted from the earth under slave-like conditions, into an actual digital device. Though many large tech corporations today have signed international conventions pledging not to use 'conflict minerals' in their hardware, the conditions under which these products are manufactured still call for concern. One of the most frequently used examples is the Taiwanese company Foxconn, which assembles a large proportion of the world's smartphones and tablets. The working conditions at Foxconn's Chinese factories (and other hardware factories) are concisely described by Nick Dyer-Witheford (2015) in his book *Cyber-Proletariat*. The word 'proletariat' was originally used by Marx to characterize the workers of

Image 5.3 Workers produce Chinese-made handsets at a factory in the Guandong province. For many years, the Guangdong province has been regarded as the sweatshop of the world.

Source: STR, Ritzau/Scanpix.

the industrial era who were free in a double sense – "free to sell their labour for a wage, but also free to starve if this sale fails" (Dyer-Witheford, 2015, p. 12) (Image 5.3).

By embracing concepts like prosumption or produsage, we risk losing sight of the fact that there is a huge difference between the *intangible* content production carried out by tech-savvy internet users in the Global North and the highly *tangible* commodity production carried out by the working poor in the Global South. It may be that, thanks to the internet, 'we are all [content] producers now', but if the notion of production is to retain any of its original meaning, we must not forget that most readers of this book (as well as its authors) will be safely positioned at the consumption-end of the global value chain, far removed from Congolese cobalt mines and Chinese assembly lines.

Production, produsage, and sociomateriality

As the attentive reader may have noticed, the Marxist argument about retaining a class-based distinction between labor and capital does not resonate particularly well with the sociomaterial credo of viewing the world symmetrically – that is, without assigning people or things certain qualities from the outset. Our argument, therefore, is not that one should simply assume *a priori* distinctions between activities or roles within the digital economy. Instead, we encourage readers to explore the affordances of digital technology in relation to organized production. One way of merging this ambition with the Marxist-inspired literature is to focus on what Braverman

(1974) calls the 'social role' of technology. Rather than assuming that automated production systems naturally reinforce the antagonistic relationship between capital and labor, or that 3D printers reverse this tendency, the point is to investigate how technology both enables and constrains certain ways of organizing production processes.

An example of how a sociomaterial perspective can be used to explore the organization of work and production in a digital context is Daniel Nyberg's (2009) study of labor processes in a call center. Based on extensive fieldwork at the call center of an Australian insurance company, Nyberg shows how networks of human and non-human actors intermingle in the construction of highly context-dependent work practices. Based on these insights, Nyberg (2009, p. 1195) concludes that "we cannot separate the roles of actors into managers/workers or people/things before analyzing activities and practices."

Notably, this does not lead him to adopt another preconceived vocabulary, in which these roles are assumed to be *conflated* rather than *separated*, which would be the case had he merely replaced the notion of production with produsage. Instead, he urges researchers to attend to the empirical messiness of local work practices and the (social) role played by technology in the constitution of these. From a sociomaterial perspective, categories such as 'producer', 'consumer', 'manager', and 'worker' should be viewed as emerging from practice rather than as categories that exist prior to local interactions. In doing so, however, one will quickly discover that the 'traditional' modes of production from industrial and even pre-industrial times often persist despite the advent of digital technology that promises to overturn such processes completely. This does not mean that labor processes today are as cruel and backbreaking as in Marx's and Taylor's age (at least not in the Global North). However, it does mean that some kind of profit-driven exploitation almost always takes place. The task for sociomaterialists interested in questions of production is to figure out exactly how digital technologies become intertwined with production or produsage processes.

? QUESTIONS

1. What characterizes the social roles of worker, manager, producer, and consumer in a given production process?
2. How is profit generated in new digitalized production processes?
3. What is the social role of digital technology in disturbing traditional production processes?
4. To what extent does digital technology give workers control over their own work?
5. To what extent does digital technology serve the purpose of specialization or deskilling?

References

Anderson, C. (2010). In the next industrial revolution, atoms are the new bits. *Wired*, 25 January. Retrieved from https://www.wired.com/2010/01/ff_newrevolution/

Barboza, D. (2016). An iPhone's journey, from the factory floor to the retail store. *The New York Times*, 29 December. Retrieved from https://www.nytimes.com/2016/12/29/technology/iphone-china-apple-stores.html

Baritz, L. (1960). *The servants of power: A history of the use of social sciences in American industry*. Middletown, CT: Wesleyan University Press.

Bauwens, M. (2005). Peer to peer and human evolution. *Integral Visioning*. Retrieved from http://z.agoravox.fr/IMG/P2PandHumanEvolV2.pdf

Bell, D. (1974). *The coming of post-industrial society: A venture in social forecasting*. New York: Harper Colophon Books.

Bird, S. E. (2011). Are we all produsers now?. *Cultural Studies*, 25(4–5), 502–516.

Bolin, G. (2011). *Value and the media: Cultural production and consumption in digital markets*. Farnham: Ashgate.

Brake, D. R. (2014). Are we all online content creators now? Web 2.0 and digital divides. *Journal of Computer-Mediated Communication*, 19(3), 591–609.

Braverman, H. (1974). *Labor and monopoly capital: The degradation of work in the twentieth century*. New York: Monthly Review Press.

Bright, J. R. (1966). The relationship of increasing automation and skill requirements. In *Report of the U.S. National Commission on technology, automation, and economic progress* (Appendix, vol. 2). Washington, DC: U.S. Government Printing Office.

Brown, R. (1992). *Understanding industrial organisations*. London: Routledge.

Bruns, A. (2008). *Blogs, Wikipedia, second life, and beyond*. New York: Peter Lang.

Burawoy, M. (1979). *Manufacturing consent: Changes in the labor process under monopoly capitalism*. Chicago, IL: The University of Chicago Press.

Castells, M. (1996). *The rise of the network society*. Oxford: Blackwell Publishers.

Comor, E. (2011). Contextualizing and critiquing the fantastic prosumer: Power, alienation and hegemony. *Critical Sociology*, 37(3), 309–327.

Cressey, P. & MacInnes, J. (1980). Voting for Ford: Industrial democracy and the control of labour. *Capital & Class*, 4(2), 5–33.

Dizard, W. P. (1981). The coming information age. *The Information Society*, 1(2), 91–112.

Dyer-Whiteford, N. (2015). *Cyber-proletariat: Global labour and the digital vortex*. London: Pluto Press.

European Parliament. (2017). Conflict minerals: The truth behind your smartphone. Retrieved from https://www.youtube.com/watch?v=yrcTxCOkuWA&t=177s

Ford, H. (1922). *My life and work*. New York: Classic House Books.

Fournier, V. & Grey, C. (2000). At the critical moment: Conditions and prospects for critical management studies. *Human Relations*, 53(1), 7–32.

Fuchs, C. (2014). *Digital labour and Karl Marx*. New York: Routledge.

Giles, J. (2005). Internet encyclopedias go head to head. *Nature*, 438(7070), 900–901.

Grossman, L. (2006). Power to the people. *TIME Magazine*, 25 December. Retrieved from http://content.time.com/time/magazine/article/0,9171,1570816,00.html

Gulick, L. (1937). Notes on the theory of organization. In L. Gulick & L. Urwick (Eds.), *Papers on the science of administration* (pp. 1–45). New York: Institute of Public Administration, Columbia University.

Gustin, S. (2011). Unpaid blogger hits 'slave owner' Huffington with $105M class action lawsuit. *Wired*, 4 December. Retrieved from https://www.wired.com/2011/04/tasini-sues-arianna/

Hamalainen, M. & Karjalainen, J. (2017). Social manufacturing: When the maker movement meets interfirm production networks. *Business Horizons*, *60*(6), 795–805.

Harman, W. W. (1974). Humanistic capitalism: Another alternative. *Journal of Humanistic Psychology*, *14*(1), 5–32.

Hoxie, R. F. (1915). *Scientific management and labor.* New York: Appleton.

Huffington, A. (2011). About that lawsuit … *Huffington Post*, 13 April. Retrieved from https://www.huffingtonpost.com/arianna-huffington/huffington-post-lawsuit_b_848942.html

Knights, D. & Willmott, H. (1989). Power and subjectivity at work: From degradation to subjugation in social relations. *Sociology*, *23*(4), 535–558.

Kumar, K. (1995). *From post-industrial to post-modern society: New theories of the contemporary world.* Oxford: Blackwell Publishers.

Littler, C. R. (1990). The labour process debate: A theoretical review 1974–1988. In D. Knights & H. Willmott (Eds.), *Labour process theory* (pp. 46–94). Basingstoke: Macmillan Press.

MacNicol, G. (2011). Here's why the unpaid bloggers suing Arianna Huffington for $105 million don't deserve a penny. *Business Insider*, 12 April. Retrieved from http://www.businessinsider.com/arianna-huffington-lawsuit-unpaid-bloggers-2011-4?r=US&IR=T&IR=T

Marglin, S. & Schor, J. (1990). *The golden age of capitalism: Reinterpreting the postwar experience.* Oxford: Clarendon Press.

Marx, K. (1978). *Economic and philosophic manuscripts of 1844.* New York: W. W. Norton & Company. (Original work published 1867.)

Marx, K. (1990). *Capital, vol. 1.* London: Penguin. (Original work published 1867.)

Masnick, M. (2011). Dumbest lawsuit ever? HuffPo sued by bloggers who agreed to work for free … but now claim they were slaves. *Techdirt*, 12 April. Retrieved from https://www.techdirt.com/articles/20110412/12162013872/dumbest-lawsuit-ever-huffpo-sued-bloggers-who-agreed-to-work-free-now-claim-they-were-slaves.shtml

McGuigan, L. & Manzerolle, V. (2014). *The audience commodity in a digital age: Revisiting a critical theory of commercial media.* New York: Peter Lang.

McLuhan, M. (1964). *Understanding media: The extensions of man.* Cambridge, MA: MIT Press.

Nogrady, B. (2016). Your old phone is full of untapped precious metals. *BBC Future*, 18 October. Retrieved from http://www.bbc.com/future/story/20161017-your-old-phone-is-full-of-precious-metals

Nonaka, I. (1991). The knowledge-creating company. *Havard Business Review*, (November-December), 96–104.

Nyberg, D. (2009). Computers, customer service operatives and cyborgs: Intra-actions in call centres. *Organization Studies*, *30*(11), 1181–1199.

Perelman, M. (2007). *The confiscation of american prosperity: From right-wing extremism and economic ideology to the next great depression.* New York: Palgrave macmillan.

Ritzer, G. & Jurgenson, N. (2010). Production, consumption, prosumption. *Journal of Consumer Culture*, *10*(1), 13–36.

Roethlisberger, F. J. & Dickson, W. J. (1939). *Management and the worker.* Cambridge, MA: Harvard University Press.

Ross, A. (2013). In search of the lost paycheck. In T. Scholz (Ed.), *Digital labor: The internet as playground and factory* (pp. 13–33). New York: Routledge.

Shapeways (2017). Create your product, build your business. Retrieved from https://www.shapeways.com

Shirky, C. (1999). RIP the consumer, 1900–1999. *Clay Shirky's writings about the Internet*. Retrieved from http://www.shirky.com/writings/herecomeseverybody/consumer.html

Siapera, E. (2017). *Understanding new media*. Thousand Oaks, CA: Sage Publications.

Smythe, D. (1977). Communications: Blindspot of Western Marxism. *Canadian Journal of Political and Social Theory*, *1*(3), 1–27.

Solow, R. M. (1957). Technical change and the aggregate production function. *The Review of Economics and Statistics*, *39*(3), 312–320.

Stempel, J. (2012). Unpaid bloggers' lawsuit versus Huffington Post tossed. *Reuters*, 30 March. Retrieved from https://www.reuters.com/article/us-aol-huffingtonpost-bloggers/unpaid-bloggers-lawsuit-versus-huffington-post-tossed-idUSBRE82T17L20120330

Tapscott, D. & Williams, A. (2006). *Wikinomics: How mass collaboration changes everything*. London: Atlantic Books.

Taylor, F. W. (1895). A piece rate system: A step toward partial solution of the labor problem. *American Society of Mechanical Engineers Transactions*, (16), 856–893.

Taylor, F. W. (1919). *Shop management*. New York: Harper. (Original work published 1903.)

Taylor, F. W. (1967). *The principles of scientific management*. New York: W. W. Norton. (Original work published 1911.)

Thompson, P. & O'Doherty, D. (2009). Perspectives on labor process theory. In M. Alvesson, T. Bridgman & H. Willmott (Eds.), *The Oxford handbook of critical management studies* (pp. 99–121). Oxford: Oxford University Press.

Toffler, A. (1980). *The third wave*. London: Pan Books.

van Dijck, J. (2009). Users like you? Theorizing agency in user-generated content. *Media, Culture & Society*, *31*(1), 41–58.

Van Dijck, J. & Nieborg, D. (2009). Wikinomics and its discontents: A critical analysis of Web 2.0 business manifestos. *New Media & Society*, *11*(5), 855–874.

Willis, P. (1977). *Learning to labour: How working class kids get working class jobs*. Farnborough: Saxon House.

Zuboff, S. (2019). *The age of surveillance capitalism: The fight for a human future at the new frontier of power*. New York: PublicAffairs.

6 COLLABORATION AND CO-CREATION

> Organization studies has always been concerned with the problem of getting individuals to collaborate toward a common goal. This chapter describes how scholars have understood collaboration among people and among organizations. From this literature, we understand that face-to-face encounters, physical proximity, and digital technologies affect how collaboration takes place in different ways. The arrival of digital technologies puts a new face on both collaboration within and collaboration among organizations. Digital technologies have also generated completely new platforms for crowdsourcing, following the co-creation and open innovation trends. This chapter argues that the increasing popularity of 'co-creation' and related concepts can be understood as a result of expectations to openness and involvement on the part of users, customers, and citizens. These expectations are afforded by advancements in digital technologies. While this chapter shows how the concepts of collaboration and co-creation differ, we will see that they are not mutually exclusive in organizational practices and analyses.

Why collaboration and co-creation?

Collaboration can take many forms, but when this joint action is systematized, planned, and formalized, it is a core component of organizations. *Within* organizations, collaboration is largely understood as the act of working with others, often practically organized in communities of interest around a common goal or shared task. For instance, if we examine the operations of media organizations, journalistic work can be observed as thoroughly collaborative: on the basis of a shared mission statement and a shared conception of their ideal readership, editors collaborate with journalists, who collaborate with each other and with their sources, to produce content (Plesner & Raviola, 2016). Collaboration can also take place *between* organizations, and we have seen such inter-organizational collaboration *among* news organizations. For instance, the International Consortium of Investigative Journalists conducted "a year-long analysis of 11.5m leaked records" on illegal financial transactions (Lewis, 2016, n.p.). The work was carried out by more than 370 journalists in 76 countries, driven by an acknowledgment of the complexity of social and political problems spanning national borders, issues that were impossible to address by any individual or single organization. As the founder of the consortium expressed it:

> while journalistic publications traditionally have been (and still are substantially) tethered to geographical locations, and other logistical local coverage considerations, information itself in the internet age suddenly was not, and is not. All of the past 26 International Consortium projects published since 2000 have involved thousands of pages of often disparate public or private records, collectively examined by leading journalists from multiple countries ... there was one, simple operating principle within the consortium: collaboration, collaboration, collaboration! (Lewis, 2016, n.p.)

The types of *collaboration* mentioned here involve a number of professionals who pursue a common project. In recent years, a different kind of collaboration has emerged: *co-creation*. Co-creation involves problem-solving and pursuit of an organization's mission involving *customers* or *users*. This is also the case in journalism, where online platforms have made it possible for media users to engage in the production of journalism and to co-create content. We see both more informal contributions such as social media content or private video footage used in mainstream journalism, and we see more formalized co-creation processes. For instance, a Finnish magazine, *Olivia*, recently launched a co-creation platform where journalists and readers collaborate to produce feature stories. A researcher who followed *Olivia*'s co-creation processes describes the setup:

> The story co-construction process is structured as a 'challenge'. Each challenge is divided into phases following standard journalistic processes of finding a story topic, angle, and interviewees; sharing experiences and expertise as content for the stories; and ideating the scene for pictures. During this early editorial stage, the magazine staff writers are in constant dialogue with the readers. With the crowd's input, journalists then write the stories. Meanwhile, the crowd can follow the story production through a progress bar on the collaboration site. (Aitamurto, 2013b, n.p.)

Organizations experiment with co-creation to create value in new ways and to increase consumer loyalty. This is also the case in the magazine example, where readers who engaged in the co-creation process felt closer to the magazine and reported that the content became more relevant. As a consequence, more readers renewed their subscriptions, and more new subscribers signed up (Aitamurto, 2013a). Co-creation not only changed relations between users and organization; it also altered organizational practices. Journalists now needed to spend much more time engaging on the online platform.

They needed to adjust their stories based on 'real readers' rather than 'ideal readers'. Journalists had a hard time balancing respect for readers' inputs and the need to respect the magazine's style, and they at times believed that the resulting stories were of a lower quality (Aitamurto, 2013a). The moral of this story seems to be that co-creation can add value, but that it demands new kinds of efforts and skills on behalf of professionals, and a recognition that the values of participants in the co-creation process are not necessarily the same. In the case of *Olivia*, co-creation works as a supplement to a traditional organization. Traditional journalistic collaboration still takes place, such that co-creation is not *replacing* collaboration, but is an add-on.

In this chapter, we discuss both collaboration and co-creation, showing how the two types of organizational practices contribute to producing value in different ways. We also argue that digital technologies afford a new type of collaboration imperative, based on a redefinition of collaboration and expertise. Hence, it is increasingly common to include laypeople in aspects of organizational decision-making. This lay participation comprises a challenge to the exclusive, protected position of experts and professionals. In the emerging co-creation paradigm, the contributors to the final product are not just organizational members with fixed positions and responsibilities; they may have unexpected origins and possess specialized, uncertified expertise. The value produced by these outsiders or amateurs is distinct from the value produced by professionals within or among organizations, and it can hardly stand alone. Therefore, both the more classical concept of collaboration and the more recent concept of co-creation are needed in analyses of digital organizing.

Collaboration as a theme in organization studies

Understood as an organized joint action to reach a common goal, collaboration is a constitutive feature of organizations. Because collaboration is ubiquitous, it is often not examined, but simply assumed to take place (Kreiner, 2008). At the same time, when it is brought to the forefront of managerial attention, it has also been called a 'management idea' and a 'fad' (Mintzberg et al., 1996). To get closer to an understanding of how organization studies have treated this phenomenon, this section first reiterates some of the economic origins of the collaboration concept. We then describe how collaboration has been discussed as a social phenomenon, for instance, as a central element of what are known as 'communities of practice' (Brown & Duguid, 1991; Lave & Wenger, 1991). The section ends with a short introduction to Computer-Supported Cooperative Work (CSCW), a tradition which early on sought to theorize the role of 'information technology' (which was a more common term than 'digital technologies' at the time) in collaboration.

Collaboration and cooperation are easily confused. It has been suggested that they can be distinguished in the following way: "Cooperative work is accomplished by the division of labour among participants, as an activity where each person is responsible for a portion of the problem solving. We focus on collaboration as the mutual engagement of participants in a coordinated effort to solve the problem together" (Roschelle & Teasley, 1995, p. 70). Cooperation and collaboration thus require different skills of employees, and accordingly, organizational choices must be made about when to cooperate and when to collaborate. When people collaborate, they need to think holistically about a given problem, whereas cooperation implies that a problem can be divided into subproblems. We can illustrate this difference with examples from the media industry. If a print medium wanted to create a revised strategy in the digital age, this would be a problem requiring collaboration among several employees. If the problem was that of composing the following day's or week's publication, this could often be solved by cooperation, each journalist and editor contributing with small elements to the whole. In the following, we focus on the problem of why and how individuals or organizations come together to solve problems, and we refer to scholars studying both collaboration and cooperation.

Economic origins

From the point of view of economic theory, one sees a tension between individuals' rational self-interest and their engagement in pursuing common goals. It is a fundamental assumption in this literature that human nature is characterized by selfishness, but that "selfishly rational actors will under certain circumstances be drawn to collaboration and cooperation" (Kreiner, 2008, p. 194). An example of such an approach is Olson's (1965) study of the dynamics of groups engaged in collective action. Olson claimed that individuals would only participate in pursuing a common goal if they directly benefited from their active engagement. If individuals have no direct incentives or if no coercion is used, Olson argued, most people will not find it in their individual interest to bear the costs of an organizational effort – they will free-ride.

Economists – and in particular behavioral economists engaged in explaining 'irrational' behavior – have since argued that altruism is more widespread than classical economists would have us believe (Kom & Ythier, 2006). In experimental economics, the problem of rational self-interest versus altruism has been tested in different ways. Ernst Fehr and Klaus Schmidt let people engage in various games, which tested selfish behavior versus altruistic collaboration. On the basis of their experiments, Fehr and Schmidt argued that under some circumstances people do make the kind of self-interested choices that conventional economic theory would expect, but in other situations they behave in a fair and cooperative manner. In their explanation of this phenomenon, they argued that cooperative behavior is dependent on the environment and on a range of social factors (Fehr & Schmidt, 1999). One might speculate whether the more compartmentalized work associated with cooperation invites more selfish behavior and whether the more holistic problem-solving associated with collaboration might lead to more altruistic behavior.

Following the economic line of thinking, collaboration can also be seen as part and parcel of competition, because even competitions (e.g. among businesses or in a game) rely on the parties cooperating about the rules of the game. In this sense, a competitive logic in society does not rule out collaboration being at the core of organizational practice. This may be an important point when considering why people engage in altruistic collaboration (or co-creation, which we will discuss later) in a competitive, capitalist economy. Collaboration may also be a way to get access to resources, and compete better, as we will discuss below.

If we move to another stream of research, namely political science, the economic conceptions of collaboration have been discussed and challenged in light of how communities actually manage to cooperate to share resources. Elinor Ostrom (1992) studied this phenomenon with a focus on how particular groups of people managed to share finite resources such as grazing land and forests. Whereas common (economic) knowledge would hold that the only way to protect finite resources is to turn them into private property, Ostrom documented that collaboration around 'the commons' has a long tradition in many societies. Ostrom was a political scientist, but on the basis of her empirical studies, she developed managerial and organizational principles for 'managing a commons'. Among these principles were: (1) define clear group boundaries, 2) ensure that those affected by the rules can participate in modifying the rules, and (3) develop a system, carried out by community members, for monitoring members' behavior. With this type of research, we move toward a view of collaboration that requires both management and refined organizational design. We begin with the design issue.

Collaboration as an organizational design issue

As already indicated, collaboration can occur both within and among organizations. These two dimensions have been referred to as intra-organizational collaboration and extra-organizational collaboration, respectively (Mintzberg et al., 1996). There is a tendency in the literature to focus on the extra-organizational collaboration – a phenomenon that has become increasingly prevalent in an increasingly complex society. As Kreiner (2008) has pointed out, the growing complexity and technological sophistication of society necessitates more intense collaboration on more occasions – resulting, for instance, in strategic alliances, joint ventures, public-private partnerships, networks, teams, and projects. Clegg and colleagues (2011) list diverse collaborations such as the Sustainable Sydney 2030 project, that engage multiple public and private stakeholders in discussions about the future of the city; or the Star Alliance global partnership allowing airlines to extend their destination choices and obtain access to each other's customers. These examples illustrate the benefits that can be achieved by making alliances, and the authors frame it as an important strategic decision whether an organization is able to pursue its goals in isolation, or whether it needs to choose one of the many forms of collaborations with partners outside the organization. In many domains, it is becoming more difficult for organizations to pursue their goals in isolation: "To the extent that the current epoch is characterized by dynamic demands on the organizations for supplementing their own resources, and by the need to share and develop knowledge resources that are hard to source in, and hard to absorb from the market, we may have an explanation for why collaboration/cooperation expands markedly" (Kreiner, 2008, p. 196). Barbara Gray made a similar observation in 1985, calling attention to the practical organizational design issues arising from the growing need to promote collaborative problem-solving across various sectors of society. Her premise was that society is marked by problems larger than any single organization acting alone can solve. She therefore argued that traditional theories of collaboration could not be applied to the new types of collaborative arrangements. Traditional bureaucratic problem-solving methods were most appropriate for routine problems and more solidly structured relationships. Another kind of thinking needed to be developed to facilitate collaboration among stakeholders who were not members of a formally established network. Gray (1985) presented a set of 'facilitative conditions' relevant to different phases of collaboration. The facilitative conditions included: (1) an identification of stakeholders, (2) an identification of mutual interdependence, (3) facilitation of direction setting, and (4) distribution of power. Recall our discussion of network organizations in Chapter 4. In theories of network organizations, focus is on the structuring of organizational entities, whereas the focus in managing collaboration is on nurturing what takes place between entities.

Scholars have argued that collaboration cannot be understood as a network composed by loosely connected individuals or organizations who share a common interest and work together to achieve common goals. Instead, collaboration often results in something resembling a traditional or formal organization. Williams and colleagues (2016) argue that the literature tends to ignore collaboration as an organizational form, and they propose that collaboration can best be understood as consisting of distinct phases which they term the 'collaboration lifecycle'. The collaboration life cycle consists of: (1) an issue phase defining the need for collaboration, (2) an assembly phase where resources and structures are agreed on, (3) a productivity phase where communication, learning, decision-making, and so on need to function well in order for the collaboration to be successful, and finally, (4) a decline phase, where collaboration is terminated or tapers off. The authors depict collaboration as a somewhat distinct organizational form, defined by phases and from such a perspective, managers are compelled to think strategically about the various stages of the collaboration lifecycle. Again, some of these ideas echo our discussions in Chapter 4, in this case our discussion of project organizations. When collaboration is in focus, however, the organizational issues revolve around definitions of common goals and processes, as illustrated in a synthesis (Figure 6.1) of the recommendations of the scholars we have discussed here.

```
Identification          Identification         Facilitation and
of issues and    →      of principles of   →   monitoring of
stakeholders            collaboration          processes
```

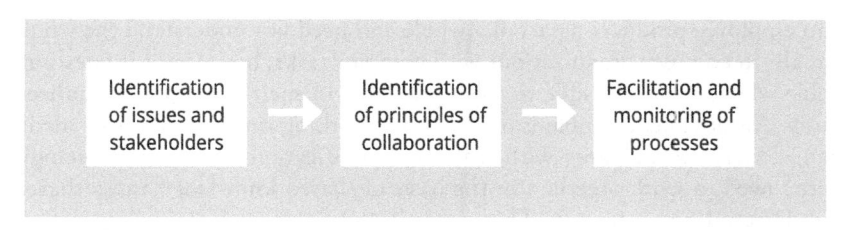

Figure 6.1 Phases in collaborative arrangements. The illustration synthesizes different contributions from the literature on collaboration.

The importance of social and physical context

We saw in the two previous sections that studies informed by economic theories place individuals' motivation for collaboration as a central concern, while organization studies have been pre-occupied with understanding collaboration as an extra-organizational phenomenon that necessitates additional organizational design considerations. Building on social constructivism, another stream of literature in organization studies focuses on the importance of the social and physical context of collaboration. This literature is critical of individualist and managerial approaches to collaboration. In the social constructivist approach, ideas about individuals cooperating to share or produce knowledge across contexts does not make a lot of sense, because social constructivists do not view solving problems as a question of motivating people to interact. The social constructivists' focus is how common social contexts are created around a shared goal.

This perspective is expressed in Jean Lave and Etienne Wenger's (1991) concepts of 'situated learning' and 'communities of practice'. In Lave and Wenger's view, knowledge does not consist of facts independent of context. Knowledge is generated by and embedded in social practices. Lave and Wenger also question some of the fundamental assumptions of behavioral economics that we discussed in the section on Economic origins. From their community of practice perspective, collaboration is not based on equal contributions of all members to a collaborative process. It may be sustained by very unequal contributions. However, all members participate in ways that are important in the reproduction and transformation of practices within the community. As an example, take the production of a movie as a collaborative project; the director and lead actors may be absolutely central to how the movie turns out, but the movie can hardly be realized without, for instance, the caterer, the runners, the lighting and sound specialists, and the extras.

In knowledge work – and much other postindustrial work – it is crucial to be able to collaborate around common goals, and this is not done by slicing up a task into small pieces and then distributing these subtasks among different persons or organizations. Such division of labor was (and is) possible in simpler work, where Fordist techniques (inspired by assembly-line production, see also Chapters 2 and 5) split up the work process into simple, repetitive operations resulting in a finished product. In such work processes, an employee produces a part of a whole and need not understand the whole at all. In complex organizations with complex tasks, however, it is questionable whether it is possible to solve problems in such a compartmentalized way. An obvious example is a laboratory working on innovation in medicine. Here, participants with different qualifications (chemistry, biology, etc.) need to work together on the basis of shared knowledge and a shared understanding of the task. This recognition has led to the following definition of collaboration: "Collaboration is a coordinated, synchronous activity that is the result of a continued attempt to construct and maintain a shared conception of a problem" (Roschelle & Teasley, 1995, p. 70).

The shared conception is heavily dependent on the establishment and maintenance of some kind of community – preferably in the same place at the same time. An example could be a laboratory, where scientists work together and have face-to-face interactions in identifying problems and deciding on solutions. In a similar vein, John Seely Brown and Paul Duguid (2000) emphasize the importance of constructing 'ecologies of knowledge'. They believe that physical proximity and informal exchange of knowledge (the proverbial water cooler/coffee machine discussions) is absolutely central to collaboration. With reference to their own work as Silicon Valley scientists, they describe how they saw collaboration being hampered when technologies allowed people to work from home, or to work together across long distances without building a vibrant, face-to-face community. The work suffered from the lack of a shared space and the kind of community-building that takes place in everyday interactions (Image 6.1).

Image 6.1 Collaboration in a knowledge-intensive environment. Most advanced innovative work is dependent on both technology and face-to-face interaction.

Source: Sanjeri, iStock.

The issue of social and physical context raises questions about the possibilities of using technologies to create collaboration across sites. These discussions are obviously central in the literature on virtual organizations, which we discussed in Chapter 4. There is also a literature dedicated specifically to technologies of collaboration. Given the explosion in software designed to afford collaboration, it is not surprising that collaborative technologies have received a great deal of attention in organization studies.

Digital technologies for collaboration

Both software applications and hardware such as smartphones and videoconferencing equipment have been used for collaboration. Such devices facilitate coordination (e.g. electronic calendars), communication (e.g. email) and storage (e.g. databases). Many online platforms (like Podio or wikis) have been designed specifically to support and streamline collaboration. However, there is nothing new about technologies fostering collaboration. James Carey (2009/1983), for instance, describes how the telegraph was the first technology to permit the effective separation of communication from transportation. This again allowed for new ways of organizing journalism, commerce, and other domains where collaboration had previously involved traveling. The telegraph also led to the establishment of standard time zones – which contributed to making collaboration smoother. Carey links the impact of the telegraph to more recent technological developments: "As the watch coordinated the industrial factory; the telegraph via the grid of time coordinated the industrial nation. Today, computer time, computer space, and computer memory, notions we dimly understand, are reworking practical consciousness, coordinating and controlling life in what we glibly call the postindustrial society" (Carey, 2009/1983, p. 21).

Among the first to investigate and theorize the role of computers in collaboration were a group of researchers assembled around the idea of Computer-Supported Cooperative Work (CSCW).

Computer-Supported Cooperative Work

In the 1980s, advances in computer technologies led researchers to begin to explore how computers could support people in their work arrangements. Recall the historical context by looking at the timeline in Chapter 1; at that point, connectivity was still at a quite primitive level, and it should come as no surprise that much collaboration using computers was shaped by what the computers could actually do. A research field formed under the umbrella term CSCW, aiming "to understand the nature and characteristics of cooperative work with the objective of designing adequate computer-based technologies" (Bannon & Schmidt, 1989, p. 3). In other words, CSCW scholars wanted to put people and their work first. At the time, this was seen as particularly pertinent because computer solutions often stood in the way of getting work done (Schmidt & Bannon, 1992). Kjeld Schmidt and Liam Bannon used the term 'cooperative' to underscore the fact that people are often mutually dependent in their work and are therefore required to cooperate to get their tasks done. Schmidt and Bannon studied the social organization of work (how tasks were allocated, how employees were rendered accountable, how employees articulated what they do), and on the basis of these studies, they sought to design systems that could support these complex work processes. They framed CSCW as the opposite of a technology-driven approach by posing questions such as:

> Why do people enter into cooperative work arrangements and how can computer-based technologies be applied to enhance their ability to do whatever it is they strive to do by cooperating? How can the coordination requirements of cooperative work arrangements be accomplished more easily, rapidly, flexibly, comprehensively, etc. with information technology? What are the implications of these requirements for the architectures of the underlying systems and services? (Schmidt & Bannon, 1992, p. 11)

In discussing digital technologies for collaboration, Michael Prilla and Carsten Ritterskamp (2010) have offered a useful distinction between CSCW, groupware, and web 2.0 applications. In line with the above, they describe CSCW as having a goal and work orientation (such as document management or intranet portals), whereas groupware supports generic tasks and has a communication and coordination orientation (e.g. instant messaging or chat). Finally, web 2.0 applications are more oriented toward playfulness and user experience (e.g. wikis or tagging communities). The authors argue that most collaboration will require elements from all types of digital collaboration support systems (Prilla & Ritterskamp, 2010). Combinations of these technological elements have been used in the design of technological platforms within organizations to facilitate collaboration – we will discuss such platforms in the coming section.

Web-based technologies for collaboration

It is clear that there has been a growing trend toward organizational uses of internet-based technologies as platforms for "social interaction, communication and collaboration" (Denyer et al., 2011, n.p.) within the organization. This type of technology use, termed Enterprise 2.0, has been heralded as a new way to deliver business benefits because of the transformative potential of more 'social', 'open', and 'participative' communication technologies. As for the organizational effects of using such platforms, it appears that they do not facilitate collaboration in any straightforward way. David Denyer and colleagues studied how the communications and IT company Telco increased its use of Enterprise 2.0 after appointing an 'Enterprise 2.0 evangelist' to the senior management team. This 'evangelist' pushed for the establishment of corporate blogging policies, the use of podcasts, and the use of organization-based wikis. The study showed that while most employees and managers agreed that Enterprise 2.0 technologies change communication practices and the level of information shared, many also reported that some issues could not be raised on these platforms, and that the issues raised there were not always substantial. The researchers call this a story of a large, high-tech, multinational company failing to make the most of the promises of Enterprise 2.0. Moreover, they did not see digital platforms contributing to improved collaboration. Collaboration technology is not a magic bullet: "The introduction of the technology alone is not sufficient to result in open collaboration and communication; rather a more dramatic change of organizational culture is needed to overcome the barriers associated with organizational politics. In order for this to be achieved, a change in leadership style may be required to one that is more collaborative" (Denyer et al., 2011, p. 392).

Besides the uses of digital technologies for collaboration *within* organizations, scholars have also discussed virtual collaboration technologies used to support collective interaction among multiple parties that are geographically dispersed. In the literature, it has been pointed out that for virtual collaboration platforms to work, several organizing activities are required. They include: (1) getting to understand the need to collaborate, (2) getting to understand the utility of the technology, (3) ensuring appropriate support for the adoption, implementation, and post-implementation phase, and (4) cultivating an organizational culture for supporting collaboration (Hossain & Wigand, 2006). In addition, collaboration in virtual environments has to rest on an understanding of the various social groups involved and on the formulation of specific common business goals that can assure linkages among technologies, organization structure, and geographical dispersion (Hossain & Wigand, 2006). There is also broad agreement among scholars of virtual teams that their success depends on how well the requirements to the digital tools are defined (Divine et al., 2011). The general point here is that considerable organizational and possibly managerial work has to be done in order to make these technologies of collaboration valuable.

In order to make most effective use of digital technologies in collaboration, some basic aspects of communication need to be considered. It has been shown that prior to engaging in virtual collaboration, initial face-to-face communication is important, just like face-to-face support is important further down the road. The relationships built through such interactions are essential to forming and maintaining trust, and to the quality of work (Hossain & Wigand, 2006). If no face-to-face interaction takes place, distributed collaboration will take place mostly in writing, thus relying on the collaborators' ability to overcome dialogical challenges in writing (Fayard & Metiu, 2014). These research-based insights show that web-based technologies may support or enable collaboration, but a range of organizational and communicative skills are required in order to make collaboration work in a digital technological environment.

Co-creation

So far, our discussions have revolved around the scholarly literature's conclusions about how collaboration can be encouraged or facilitated. This applies at the level of individual employees (who have to see the point in collaborating), at the level of teams, and among organizations as a whole. We saw how different digital technologies have been developed and discussed as platforms for such collaboration, but we have not yet shown how digital technologies are believed to afford value creation *together with customers, users, or citizens*. When such collaboration takes place with the participation of organizational 'outsiders', it is known as 'co-creation'. The concept of co-creation has many parallels to that of produsage (which we discussed in Chapter 5), since both rely on the participation of people who have no formal obligations to share ideas or interact with the organization.

Although co-creation is also a topic of organizational studies, it has received particular attention in the neighboring strategy literature. What is particular about co-creation, judging from this literature, is that it is starting to become a strategic necessity. The need for co-creation arises because it is getting increasingly difficult to create value in a company-centric way. This logic is similar to what we will discuss later on in relation to communication. Our argument there is that people have gotten used to being active and involved in dialogues rather than addressed through one-way communication from an authoritative sender (see Chapter 8). Co-creation is interesting in relation to our discussion of free-riding in the beginning of this chapter, since the idea that individuals will free-ride whenever it is not in their interest to collaborate seems to contradict the logic of co-creation. The operative assumption in the co-creation literature is that individuals *want* to be engaged, and that their loyalty will shift to competitors if organizations are not able to find ways to engage them.

Among both managers and scholars, co-creation is thus constructed as a response to changes in the environment. Strategy scholars C. K. Prahalad and Venkat Ramaswamy (2004a) are proponents of co-creation as a preferable way to produce value. They argue that it is no longer just the role of *companies* to create value. Value creation is an option for *consumers* as well. Consumers have become "armed with new tools and dissatisfied with available choices" (Prahalad & Ramaswamy, 2004b, p. 4), so they want to interact with companies to co-create products. According to the authors, consumers now insist on being connected, informed, and active – rather than isolated, unaware, and passive. It is obviously a strong claim that consumers have this urge in present-day society, but in any case, it is observable that co-creation is a growing trend (Image 6.2).

The openness paradigm

This co-creation diagnosis comes with new organizational challenges. If businesses want to become 'customer-centered', they need to engage with

Image 6.2 The façade of the NYSE in 2012 on the occasion of the public offering of the review app YELP. With user reviews of a multitude of businesses and services, YELP is but one example of the growing number of review sites that rely on crowdsourcing.

Source: Richard B. Levine, Ritzau/Scanpix.

customers at every point where value is created: "not just sales, marketing and service, but also product design, supply chain management, human resources, IT and finance ... Customer interaction in these areas often leads to open collaboration that accelerates innovation using online communities" (Berman, 2012, p. 20). This quote links co-creation to what we might call the 'openness paradigm' (which we also briefly discussed in Chapter 1). Among managers and scholars alike, there is a growing interest in how organizations can 'open up' to create value in new ways: through 'open innovation' (Chesbrough & Appleyard, 2007) and 'open strategy' (Whittington et al., 2011), or as we see in the quote, 'open collaboration'. All these partly related concepts derive from open-source software communities (relying on experts) and user innovation (relying on users) (von Hippel, 1986). Innovation processes and strategy processes have traditionally been kept inside organizations. They relied partly on opacity because of fear of imitation and loss of competitive advantage. Ideals about openness lead organizations to become more inclusive and to increase transparency beyond the boundaries of the firm.

Many scholars link the new expectations to organizations with digital technologies. Consider this quote: "most organizations will need to factor in the impact of information technology and the managerial advantages of openness, both internally and externally. To a large extent, greater external transparency is probably unavoidable" (Whittington et al., 2011, p. 540). In relation to our larger topic of digital organizing, this discourse indicates a more or less uncontested view that digital technologies automatically lead to public demands for openness, involvement, and transparency. This also means that although the technologies do not in themselves force organizations to do anything, their affordances create expectations about particular organizational behaviors. We discuss the transparency demands that organizations face in more depth in Chapter 9. In the following section, we will focus on a case showing the type of public expectations organizations may meet, the organizational efforts needed to create involvement, and the role of digital technologies in this work.

In the co-creation paradigm, organizations need not only understand their relationship to customers, clients, or users differently, they also need to develop new practices and new competencies. As Ramaswamy remarks:

> the components of value creation now entail global resource networks of partnering firms and suppliers as well as communities of individuals (customers) outside the firm. To grow in this environment, companies must have the competence to manage and influence large numbers of collaborators – including skilled individuals, communities of customers, and many varieties of stakeholders around the globe. This de-centering and democratization of the process of value creation is spreading into many industries. (Ramaswamy, 2008, p. 11)

Ramaswamy ties the whole co-creation trend to the development of internet technologies, notably search engines, engagement platforms such as review sites, internet-based interest groups, high-bandwidth communication, and social interaction technologies.

The degree to which co-creation is tied to digital technologies is open to debate, but in Ramaswamy's view, such technologies have empowered customers like never before. Let us examine some examples of how digital platforms afford co-creation practices and how different types of organizations try to leverage their affordances.

Digital platforms for co-creation: The case of Nike

If we look at large, private organizations, Nike is a commonly used example of how a company can co-create value with customers. It has involved customers in the value-creation process through the establishment of internet sites where customers can customize products, provide ideas on how to improve products, and so on. The company has several co-creation projects. In Nike's running shoe business, for example, a new project, NikePlus, was launched in 2006. NikePlus was based partly on a collaboration between Nike and Apple. It consisted of an Apple iPod music player, a wireless device to connect the music player to running shoes, Nike shoes with a pocket for the wireless device, and memberships in online communities (itunes.com and nikeplus.com) (Ramaswamy, 2008). So, what was the value co-created with customers here? The tracking possibilities allowed Nike to learn directly from the behavior of customers and obtain direct input from them. On the customer side, they allowed people to track their exercise (running) with unprecedented precision, to take part in new social networks, to integrate running and music, to make and state running goals easier, and so on.

According to Ramaswamy, the customer involvement allowed Nike to reduce the costs of marketing through positive word of mouth and to reduce the risk of customer dissatisfaction. Customers allegedly felt a better price-experience relationship (Ramaswamy, 2008). In practice, though, what seemed to be a fairy tale of co-creation of value has also proven to demand continuous involvement and communication on the part of the company. For example, in 2016, Nike made a new version of their app and received an avalanche of criticism on their Facebook page. As a tech commentator put it: "Despite a tagline claiming the new version was made 'for runners, by runners', many old Nike+ users are livid with the sudden switch and the loss of functionality and features that made Nike+ Running feel so vital" (Welch, 2016, n.p.).

Nike had to engage actively in the monitoring, communication, and adjustments of the product. A few examples show how the company tries to keep its promise of involvement with the users and how passionate users seem to be about the functionalities of the product. It seems that the company cannot just develop and design a product as they see fit:

> Nike writes: "Cynthia – Help is here. Once you have completed a run, you'll have to share it to the feed. Then it will ask you to share again and then you can select which social media you'd like to share your run on."
>
> A new user replies: "I don't want to share anything to any feed, only post to my closed Facebook and closed Instagram accounts. Not post to a pseudo wanna-be social platform I have not asked for or know anything about, including the privacy policy."
>
> In another example, Nike writes: "Brigitte – your previous coach program is not available in the new app, but we've got you covered. Check out one of our race-day training plans to pick up where you left off. We recommend you choosing a plan, then adapting it to fit your schedule."
>
> Here, another new user replies: "Stop saying that crap ... People ran for months on that training plan. Had achievements they were proud of and worked hard for, all to be taken away by an update which is nonsense" (all quoted by Welch, 2016, n.p.).

The critical comments not only appeared on Nike's Facebook page, but managed to flood the app stores and various tech discussion forums. This is not surprising, given that the app has millions of users. As a consequence, Nike seemed to be working overtime to respond to the criticism and could only hope that users would not migrate to competing apps (Welch, 2016). The case tells us several things about co-creation as an organizational phenomenon: First, that it is not enough to 'open up' in an early phase of product development; continued dialogue with users is needed. Second, that communication with users cannot be contained on designated digital platforms; it spills into an uncontrollable number of websites. And third, that people are not only eager to engage, they may also become furious if they are left out of decision-making processes.

Co-creation is also used in the public sector, where the above-mentioned demands for openness and opportunities for participation are met with an increased focus on innovation involving citizens. Citizens are assumed to have an interest in raising the quality of public service, so they are invited into both process innovation and product innovation, just as they are asked to participate in making sense of public data that was hitherto kept inside public organizations. In the section on Collaboration, co-creation and sociomateriality, we discuss an example of this public engagement in some depth. Here we will examine how digital crowdsourcing platforms are conceived of as particularly conducive to co-creation practices.

Organizing crowds digitally: The case of Innosite

Online platforms for 'crowdsourcing' (Howe, 2009) offer many instances of people who are external to an organization, but engaged in producing value for it. Kamstrup (2017, p. 15) defines crowdsourcing as involving a central task giver, people engaged in the task, online interactions between people, and a compensation structure. People are not always compensated with money, however. They can receive no reward at all; they can be rewarded with a minor sum for each task; or the setup can be competitive, such that only the winner gets a reward (Kamstrup, 2017, p. 90). As seen here, crowdsourcing reintroduces the issue of incentives that we discussed on the basis of economic theory, but individuals now participate without necessarily being rewarded, as we would expect from the literature on collaboration. Kamstrup explains how the reward structures can differ:

> Some offer substantial monetary rewards (i.e. Xprize, 2017; Innocentive, 2017), while others offer the honour of winning and the possibility of having the winning proposal realised (i.e. openIDEO, 2017). Some initiatives identify themselves as marketing campaigns for large companies (i.e. Starbucks, 2017), whereas others are presented as 'amateur' science projects (i.e. eBirds, 2017). Some demand highly technical and profession-specific inputs (Netflix, 2017), and others are designed so that virtually anyone can participate (i.e. Lego, 2017). (Kamstrup, 2017, p. 41)

The proliferation of digital crowdsourcing platforms testifies to the fact that it is indeed possible for organizations to engage people in their problem-solving and value creation. There seems to be interest enough from users, customers, and citizens. At the same time, the organizational efforts required to make co-creation work are significant, and a number of decisions need to be made about how to create value (Table 6.1). We see this in Kamstrup's (2017) study of a platform designed for co-creation of innovative solutions to the building industry. Kamstrup followed the creation of the platform, which was inspired by the open innovation paradigm as well as notions of collaboration and co-creation. The platform's operating team and the design company behind it talked about co-creation as an important factor for the platform's potential for success. The idea is presented on the platform's website as follows:

> Innosite builds on the idea of open innovation which means that companies involve external players and knowledge in their development processes. In this way ideas, problems and solutions are taken beyond their usual subject- and organisation-specific contexts. Open innovation platforms facilitate the involvement of users and experts in the development processes. This is because sharing, collecting and selecting ideas and solutions can be done both cheaper and faster than in traditional development and, moreover, independently of the individual project. (Kamstrup, 2017, p. 42)

To realize the platform, funding was provided from large private actors in the building industry, and specific challenges or 'competitions' were

Table 6.1 Dimensions of digital platforms for collaboration (Kamstrup, 2017). The table shows how platforms differ in terms of who has access to them, how the crowd is put together, how long it has been active, how much information is shared, and whether it is designed for collaboration or competition. Organizational decisions need to be made about each dimension.

Dimensions	Extremes	
Access	Open	Invited
Crowd composition	Diverse	Uniform
Permanence	Single	Multi
Level of information	Full access	Limited access
Interactions among crowd members	Collaborative	Competitive

formulated together with actors from the industry. A lot of work had to be done to assemble 'the crowd' (which is not just 'out there'), a lot of work had to be done to design the platform, and a lot of work had to be done to make the competition process run smoothly. As Kamstrup describes:

> some crowd members experienced technical issues when trying to upload their proposal, however when asked about the more 'soft' sides of participation there is a common agreement that it is easy to participate. The question in the competition brief is formulated in an understandable and not too technical manner and in general the community manager takes time to clear out misunderstandings about what the initial question is about. (Kamstrup, 2017, p. 123)

The role of the platform manager is crucial to making things work, and so is the collaboration between actors around the platform. Open innovation does not just fall out of the sky. We previously discussed the 'openness paradigm'. In this case, the platform designers tried to build openness into the platform by trying to reach far with the competition brief and by encouraging openness *between* crowd members. In the organization, it was stated that for the platform to "really become a success, co-creation between random crowd members was needed" (Kamstrup, 2017, p. 124). This meant that the platform was designed to make a proposal visible to the entire crowd as soon as it was uploaded. In practice, this 'openness' meant that proposals became rather standardized. Mostly, participants would keep their ideas to themselves until the competition closed, but the ideas that had previously been successful and were searchable on the platform were to some degree imitated later on. So instead of affording innovative proposals, the platform tended to afford more standardization (see also Kamstrup & Husted, forthcoming). The design of the digital platform, the construction of 'the crowd', and the management of the co-creation processes were organizational tasks that proved crucial to the value created in the end. The case exemplifies the relevance of discussions about 'the wisdom of the crowd' (Surowiecki, 2005) versus 'the stupidity of the crowd' (Le Bon, 2002/1895). Sometimes more people can produce more valuable insights; at other times, the way the crowd is assembled is problematic or the crowd's input is useless (see for example, Barrett, 2018).

Collaboration, co-creation, and sociomateriality

In this chapter, we examined how organization studies have been concerned with the problem of motivating individuals to collaborate toward a common goal, and how scholars have understood collaboration among people and among organizations. We have also discussed the role of face-to-face encounters and physical proximity, as well as the role of digital technologies in collaboration. Summarizing the insights produced by organization studies and discussed throughout this chapter, it is fair to say that collaboration can be analyzed as a sociomaterial phenomenon that depends on both the people who participate and the technological/material conditions in which they are embedded. The specific designs of digital technologies and the specific social interactions around them imply that the technologies' affordances for collaboration vary and the organizational implications are manifold.

We also discussed co-creation as a result of expectations to openness and involvement on the part of users, customers, and citizens and as afforded by advancements in digital technologies. The concepts of collaboration and co-creation differ, but they are not mutually exclusive in organizational practices. We can illustrate this with an example from the Danish public sector, where the Agency for Culture, together with the major public cultural institutions under the agency, experimented with co-creation through a series of hackathons. A hackathon is a sprint-like event where computer programmers, designers, and subject-matter experts come together to create usable software.

In this case, we see how various institutions come together to collaborate around a co-creation project, with the dual aim of making enormous data sets more valuable and of encouraging citizens to become partners in bringing the national cultural heritage alive. One of the hackathons was designed in collaboration between the National Museum, The Royal Library, and the Agency for Culture. The collaboration consisted of defining the goal of the project, assembling data sets, and making them public, arranging a two-day physical event (with ignite talks, logistics, and selection of a winner), communicating results, and so on (version2, 2012). The co-creation aspect was part of the project because the public institutions worked on 'opening up their data' and allowing anyone access to make use of them (#HACK4DK, 2019). The process was thus not only about defining a specific goal and persuading people to collaborate, but also about inviting people who were not formally attached to the participating organizations into particular projects and allowing them to set directions – hoping to create value in creative and surprising ways. Commenting on the value created in this particular project, one of the co-organizers wrote in an IT-forum: "We really benefited from your participation. Our data got attention, we got good ideas on how to use them and lots of feedback on how to manage them" (version2, 2012, our translation). A participant defined the value in this way: "Super nice ambience and lots of conversations across institutions and professional domains. Really inspiring, so thumbs up to the hosts" (version2, 2012, our translation).

In this case, the digital component in the organization consisted of data sets around which participants collaborated with the organization. In other cases, the digital component may be a particular platform or something else – but this does not change the fact that digital organizing depends as much on organizing as on digital technologies. The need to organize digital collaboration and co-creation is no exception. When we analyze such organizational practices, it is worth being aware of the promises and pitfalls of the rather fashionable terms of 'co-creation', 'crowdsourcing', and so on. Here again, we need to hold them up against the classical concepts (and organizational aims) of collaboration and cooperation. The latter may produce insights into some of the purposes and dynamics of the former.

> **? QUESTIONS**
>
> 1. To what extent is collaboration taken for granted or worked with strategically in given organizations?
> 2. What facilitates or hinders collaboration?
> 3. Which roles do materiality and digital technologies play in collaboration?
> 4. How do organizations experience the expectations to openness and involvement on the part of citizens, users, or customers?
> 5. How is co-creation understood and practiced?
> 6. (How) does the openness paradigm influence how specific organizations think about collaboration and co-creation?
> 7. How is openness organized?

References

#HACK4DK (2019). Hack your heritage! Retrieved from https://hack4.dk

Aitamurto, T. (2013a). Balancing between open and closed. *Digital Journalism*, *1*(2), 229–251.

Aitamurto, T. (2013b). The challenges of co-creating magazine journalism with reader input. Retrieved from http://mediashift.org/2013/06/challenges-cocreating-magazine-journalism-reader-input/

Bannon, L. J. & Schmidt, K. (1989). CSCW – Four characters in search of a context. *DAIMI Report Series*, *18*(289), 1–20.

Barrett, H. (2018, February 11). Remember the stupidity of crowds. *Financial Times*. Retrieved from https://www.ft.com/content/d0835352-0b66-11e8-839d-41ca06376bf2

Berman, S. J. (2012). Digital transformation: Opportunities to create new business models. *Strategy & Leadership Journal of Business Strategy*, *40*(4), 16–24.

Brown, J. S. & Duguid, P. (1991). Organizational learning and communities-of-practice: Toward a unified view of working, learning, and innovation. *Organization Science*, *2*(1), 40–57.

Brown, J. S. & Duguid, P. (2000). *The social life of information*. Cambridge, MA: Harvard Business School Press.

Carey, J. W. (2009). Technology and ideology: The case of the telegraph. In *The communication & media studies collection* (pp. 1–21). London: SAGE. (Original work published 1983.)

Chesbrough, H. W. & Appleyard, M. M. (2007). Open innovation and strategy. *California Management Review, 50*(1), 57–76.

Clegg, S., Carter, C., Kornberger, M. & Schweitzer, J. (2011). *Strategy: Theory and practice*. London: SAGE.

Denyer, D., Parry, E. & Flowers, P. (2011). 'Social', 'open' and 'paricipative'? Exploring personal experiences and organisational effects of enterprise 2.0. *Long Range Planning, 44*(5–6), 375–396.

Divine, M., Schumacher, M. & Cardinal, J. S. L. (2011). Learning virtual teams: How to design a set of Web 2.0 tools?. *International Journal of Technology Management, 55*(3/4), 297–308.

Fayard, A.-L. & Metiu, A. (2014). The role of writing in distributed collaboration. *Organization Science, 25*(5), 1391–1413.

Fehr, E. & Schmidt, K. M. (1999). A theory of fairness, competition, and cooperation. *Quarterly Journal of Economics, 114*(3), 817–868.

Gray, B. (1985). Conditions facilitating interorganizational collaboration. *Human Relations, 38*(10), 911–936.

Hossain, L. & Wigand, R. T. (2006). ICT enabled virtual collaboration through trust. *Journal of Computer-Mediated Communication, 10*(1), n.p.

Howe, J. (2009). *Crowdsourcing: Why the power of the crowd is driving the future of business*. New York: Three Rivers Press.

Kamstrup, A. (2017). *Crowdsourcing and the architectural competition as organisational technologies*. Frederiksberg: Copenhagen Business School.

Kamstrup, A. & Husted, E. (forthcoming). Is crowdsourcing anarchist? Interrogating the libertarian spirit of two online platforms. In M. Parker, T. Swann & K. Stoborod (Eds.), *Anarchism, organization and management*. London: Routledge.

Kom, S.-C. & Ythier, J. M. (2006). *Handbook of the economics of giving, altruism and reciprocity: Foundations*. Amsterdam: Elsevier.

Kreiner, K. (2008). Collaboration and cooperation. In S. Clegg & J. Bailey (Eds.), *International encyclopedia of organization studies* (pp. 194–203). Los Angeles, CA: SAGE.

Lave, J. & Wenger, E. (1991). *Situated learning: Legitimate peripheral participation*. Cambridge: Cambridge University Press.

Le Bon, G. (2002). *The crowd: A study of the popular mind*. Mineola, NY: Dover Publications. (Original work published 1895.)

Lewis, C. (2016). The future of journalism in three words: Collaboration, collaboration, collaboration. *The Guardian*, April 18. Retrieved from https://www.theguardian.com/commentisfree/2016/apr/18/future-of-journalism-collaboration-panama-papers

Mintzberg, H., Jorgensen, J., Dougherty, D. & Westley, F. (1996). Some surprising things about collaboration: Knowing how people connect makes it work better. *Organizational Dynamics, 25*(1), 60–71.

Olson, M. (1965). *The logic of collective action: Public good and the theory of groups*. Cambridge, MA: Harvard University Press.

Ostrom, E. (1992). Governing the commons: The evolution of institutions for collective action. *Natural Resources Journal, 32*(2), 415–417.

Plesner, U. & Raviola, E. (2016). Digital technologies and a changing profession. *Journal of Organizational Change Management, 29*(7), 1044–1065.

Prahalad, C. K. & Ramaswamy, V. (2004a). Co-creation experiences: The next practice in value creation. *Journal of Interactive Marketing, 18*(3), 5–14.

Prahalad, C. K. & Ramaswamy, V. (2004b). Co-creating unique value with customers. *Strategy & Leadership, 32*(3), 4–9.

Prilla, M. & Ritterskamp, C. (2010). The Interplay of Web 2.0 and collaboration support systems: Leveraging synergies. In D. Randall & P. Salambier (Eds.), *From CSCW to Web 2.0: European developments in collaborative design* (pp. 193–218). London: Springer.

Ramaswamy, V. (2008). Co-creating value through customers' experiences: The Nike case. *Strategy & Leadership, 36*(5), 9–14.

Roschelle, J. & Teasley, S. D. (1995). The construction of shared knowledge in collaborative problem solving. In C. O'Malley (Ed.), *Computer supported collaborative learning* (pp. 69–97). Berlin Heidelberg: Springer Verlag.

Schmidt, K. & Bannon, L. (1992). Taking CSCW seriously supporting articulation work. *Computer Supported Cooperative Work (CSCW), 1*(7), 7–40.

Surowiecki, J. (2005). *The wisdom of crowds*. New York: Anchor Books.

version2 (2012). Udviklere samlet til kulturarvshackathon ved #hack4dk. *Version 2.* Retrieved from https://www.version2.dk/blog/udviklere-samlet-til-kulturarvshackathon-ved-hack4dk-48167

von Hippel, E. (1986). Lead users: A source of novel product concepts. *Management Science, 32*(7), 791–805.

Welch, C. (2016). Nike redesigned its popular running app, and users are very angry. *The Verge*, April 27. Retrieved from https://www.theverge.com/2016/8/27/12670716/nike-running-app-bad-redesign

Whittington, R., Cailluet, L. & Yakis-Douglas, B. (2011). Opening strategy: Evolution of a precarious profession. *British Journal of Management, 22*(3), 531–544.

Williams, C. M., Merriman, C. & Morris, J. C. (2016). A life-cycle model of collaboration. In J. C. Morris & K. Miller-Stevens (Eds.), *Advancing collaboration theory: Models, typologies, and evidence* (pp. 175–196). New York: Routledge.

7 KNOWLEDGE AND DATAFICATION

> In organizations – and in organization studies – knowledge and data are considered resources that are crucial for organizations to work smarter, deliver higher quality, and/or gain competitive advantage. Both in practice and in the literature, knowledge and data are often portrayed as something organizations possess and should be better at identifying and leveraging, or alternatively as something 'out there', which organizations should try to obtain. Knowledge has a longer history than data as a matter of concern for organizations, and the discipline of knowledge management has developed various techniques for 'capturing' and sharing knowledge. In this tradition, people are often the center of concern, as they are considered bearers of tacit knowledge. While organizations' concern with knowledge management has intensified over several decades, advancements in digital technologies have created new conditions for these practices. The new technologies were originally conceived as simple repositories for saving and sharing information. This function has given way, however, to the production of knowledge becoming intertwined with new ways of using and handling data – for instance through the use of artificial intelligence (AI) or automated treatment of large data sets. Data has acquired another status – it is no longer lowest in the hierarchy, below 'information' and 'knowledge', but seen as having an agency of its own. In this chapter, we speak of datafication to point to this qualitatively different knowledge paradigm in organizations. We not only discuss how both 'knowledge' and 'datafication' are considered crucial resources to be harvested and managed, but also how these concepts differ in their relationship to people and technology.

Why knowledge and datafication?

The conventional understanding of the 'knowledge society' is that it is the culmination of the vast increases in data creation and information dissemination resulting from innovation in information technologies (Välimaa & Hoffman, 2008). However, knowledge and the valuation of knowledge are not simply an effect of technological advances. To understand how knowledge has become a valuable asset not only at the societal level but at the organizational level, we need to examine how knowledge is conceived, produced, activated, and deployed in organizations. In the organizational literature, knowledge has often been portrayed as a predominantly human phenomenon, but also as intertwined with the use of technologies. This chapter will examine how the literature helps us understand knowledge as a product of the interplay between technologies and people. In the early days of knowledge management, the focus was on the technological systems that allowed for the storage and sharing of information. However, these systems were always seen as dependent on people. Similarly, organizations experimenting with new forms of knowledge such as AI and machine learning, remain highly dependent on human decisions and strategies for knowledge production and knowledge sharing. Although the themes of knowledge and datafication raise different types of questions, many of these questions are about the relationship between people and technologies in organizations.

Knowledge and datafication in practice: The case of home care

Let us look into a case where the issue of knowledge is critical, and where the 'management' of employees' knowledge is intertwined with the use of different kinds of technologies. In home care organizations, managers are highly concerned with issues of documenting the work of home care employees and the citizens' health conditions. Home care employees drive from home to home, assisting elderly or disabled citizens with bathing, eating, and medicine dosage. Since employees operate in shifts and rarely see their colleagues, they need reliable methods to exchange knowledge, and since they are responsible for administering the proper treatment, they are dependent on accurate medical information. The quality of the care they provide is crucially dependent on how knowledge about their clients is produced, stored, used, and exchanged. In such organizational settings, many factors complicate knowledge management tasks, even if we normally regard these kinds of service organizational settings as not being knowledge intensive. For instance, some home care workers have little or no higher education and some have difficulties reading and writing. Also, driving from client to client on a tight schedule, employees have no common workspace and are pressed for time.

In a municipal home care organization in a large city,[1] managers sought to compensate for these factors by producing manuals, having employees take notes from every visit, and having employees return to the main office at the end of their shift to enter information into a shared database. However, such practices are challenged by, for instance, the lack of a shared terminology and the perceived tediousness of documenting all the information from employees' home care visits. With handheld devices, new

---

1 This case description is based on an interview with a manager in a municipality. It is part of Plesner and Justesen's ongoing research on the digitalization of public sector organizations.

possibilities for documentation (and knowledge management) have arisen. In this particular organization, a project was initiated to register the work processes involved in different types of visits, and to digitize knowledge production and knowledge exchange. All care workers received a handheld device with a special app installed. This app generates push notifications in the form of red alerts if new medicine standards are introduced or new dosages prescribed. The 'red alert' frees employees from consulting catalogues and prescriptions. The app also has drop-down menus where employees can enter their assessment of the client's condition without having to write a long notation. This partially solves the problem of employees who have trouble reading and writing.

The fact that the device is portable addresses and partially solves other problems as well, namely the problem of the distributed organization and the employees' waiting time to enter data in the shared computer in the main office. But with this new knowledge management project, new issues arise in the organization. System designers must ensure that the drop-down menus have the correct options, that the hardware, software, and push information work smoothly, that the home care aides do not feel even more detached from their organization, and that the employees use the technology as intended. Also, the issue of bodily knowledge and tacit knowledge is left unaddressed; how are subjective experiences and know-how communicated? Finally, issues of surveillance arise because managers are now able to track employees and their work in a much more detailed manner, literally minute by minute. We mentioned that datafication basically means that data acquires a new status and centrality in organizations (and beyond). This is also the case in this home care organization. Data acquires a new role as a management tool, both because more data is made readily available for employees, because employees produce more standardized information that can be analyzed, and because management has new types of data available on which to base their management decisions (Image 7.1).

As we see from this case, the use of knowledge in organizations is related to the data shared among or generated by employees. But the relationship between knowledge and data is not a simple, one-to-one relation. It is not the case that 'the more data we have, the more knowledge we have'.

In this chapter, we discuss how knowledge has become established as an important asset for organizations by management scholars in organization studies and how knowledge management has evolved as a practical discipline that can support organizations in eliciting and sharing knowledge. We highlight the basic assumptions and technologies of knowledge management. We then show how knowledge and expertise in organizations take on new forms as digital technologies produce new conditions for storing, sharing, learning, and connecting. We focus here on datafication, which we will discuss as a managerial concern with using data more extensively and in new ways in organizational contexts. The twin concepts of knowledge and datafication help us understand the relation between concerns with knowledge and concerns about data and its use. Managers consider both

Image 7.1 Handheld devices have become more common in home care. Many home care workers are encouraged to engage their clients when working with the technology. The idea is that screen time has the potential to become quality time if clients are involved in the planning of their treatment and care.

Source: Shapecharge, iStock.

knowledge and data to be resources to make work smarter, deliver higher quality, and/or gain competitive advantage. Still, we will treat knowledge and data separately here because the two words point to the role of technology in different ways; technology is primarily an instrument in the knowledge management tradition while it plays a more central role in the datafication paradigm.

Knowledge as a theme in organization studies

Post-industrial societies (Bell, 1974) are marked by an increasing valuation of knowledge. Not only are growing tertiary and quaternary sectors (i.e. the service sector and the information-based service sector) essentially knowledge-oriented, but knowledge is also increasingly seen as a form of capital that adds value to other forms of capital. As one management scholar puts it: "Along with the traditional resources of land, labor and capital, knowledge has always been important in determining both a firm's and a nation's competitive success. The 'know-how' of selected employees has always been a valuable asset" (Cole, 1998, p. 15).

As knowledge began to be portrayed as an economically important asset to organizations, it also began to attract interest from organization scholars. Initially, there was a relatively tight linkage between 'knowledge' and the individual employee. Individual employees were seen as embodying knowledge and expertise stemming from experience. Hence, Peter Drucker (1957) coined the term 'knowledge worker' to account for a new and important type of employee (Blackler, 1995). In the context of innovation, it became a key goal to motivate knowledge workers to leverage the organization's knowledge and share knowledge with one another to foster innovation. In this line of thinking, it is paramount to make the best possible use of individuals' innovative capacities and make it collective, or 'de-individualize' it. As long as knowledge is conceived as attached to individuals, the organizational problem becomes one of extracting and amplifying this knowledge and then incorporating it into organizational practices. Only in this way can individual knowledge be elevated into *organizational* knowledge (Nonaka, 1994).

Personal or experience-based knowledge is often referred to as 'tacit knowledge'. The concept is widely used in organization studies, with reference to a philosopher of science, Michael Polanyi. Polanyi famously stated that "we can know more than we can tell" (Polanyi, 1967, p. 4), trying to conceptualize how logical reasoning does not stand alone in the development of knowledge. The latter is infused with personal knowledge, informed guesses, and imaginings based on an individual's experience. This wider view of knowledge implies that the ensemble of employees and their tacit knowledge can become a key asset in a company's competitive advantage, because such knowledge-assets are notoriously hard to account for and thus copy. As Robert E. Cole (1998, p. 17) observes, "many non-tradable assets arguably have critical strategic value precisely because they are not easily imitated or traded". This view of knowledge can help explain why a loss of experienced employees or a high turnover can be problematic for an organization. It has also inspired a whole field of research to interrogate how individual employees' knowledge can be turned into organizational learning.

Organizational learning

Organizational learning has been described as the goal of knowledge management (King, 2009, p. 5) or as a process of "encoding influences from history into routines that guide behavior" (Levitt & March, 1988, p. 319). The organizational learning literature focuses on how social groups, teams, and networks construct and communicate knowledge – for instance, in communities of practice (Lave & Wenger, 1991) or in practices beyond their work communities (Brown & Duguid, 2001). The issue of establishing new and improved work routines lies at the core of various management systems such as 'continuous improvement processes' or 'lean management'. These systems offer guidelines to identify opportunities for becoming more efficient, eliminating waste, adding value, and so on.

Just like knowledge in organizations has been conceived as a largely human or social phenomenon, so has learning. Although machine learning has become an issue for some organizations, and is sometimes coupled to organizational learning (Hasebrook & Maurer, 2004; Lant & Shapira, 2001), human and machine learning are often disconnected in practice and in theory.

The value of human knowledge and learning: The case of Watson

Discussions of the value of human knowledge and learning have been invigorated in the context of experiments with AI (see also our discussion in Chapter 1). In the area of healthcare and medicine, for instance, IBM's flagship project, Watson, was initially built to compete with humans in the game Jeopardy,[2] but has since been developed to serve various other purposes, like assisting doctors in making cancer diagnoses. Immense computing power allows Watson to quickly assemble data from all available studies in a given field and to suggest diagnoses. The machine needs to be 'trained' by individuals giving feedback regarding its suggestions. In some areas such as image scanning, Watson seems to be a valuable support in decision-making, but it obviously cannot replace human judgment or the hunches and personal experiences of professionals.

In the burgeoning field of AI, intense discussions are taking place regarding the relationship between individuals' knowledge and digital technologies' capacity to provide *useful knowledge* – understood as something other than mere information. For instance, critics point out that despite claims to the contrary, Watson cannot generate new insights, and that it is a struggle to update Watson's knowledge base. It has taken almost six years for data engineers and doctors to train Watson to identify and then make decisions about seven types of cancer, and they still struggle to update the system (Ross & Swetlitz, 2017). As a 'lead Watson trainer' puts it: "Changing the system of cognitive computing doesn't turn around on a dime like that – you have to put in the literature, you have to put in cases" (Ross & Swetlitz, 2017, n.p.). For instance, a research presentation at a cancer conference recently led to changed treatment guidelines for one type of cancer within a week, and Watson had to be trained accordingly afterwards; it could not by itself identify this new phenomenon at such a speed. Watson relies on scientific articles which have a long publication process. It takes a long time before the literature shows a paradigm shift, so cutting-edge research results had to be picked up by humans. As we can see, AI re-actualizes the question of human knowledge and learning in organizations. It is also central to discussions of data-driven decision-making and datafication, which we discuss in more detail in the second half of this chapter, in the section on Datafication. For now, we will dwell on some 'pre-AI' technologies to handle knowledge.

---

2 Jeopardy was originally an American TV quiz show game where contestants are given the answers and have to guess the question. The game is thus a form of logical challenge and tests contestants' powers of association, inference, and logic.

Knowledge management

For managers – and organizational scholars – the ambition to make use of employees' knowledge poses a practical challenge because some knowledge is difficult to measure or difficult to make visible and tangible. Tacit knowledge is laboriously developed over time, and from the point of view of the knowledge management literature, it tends to be underutilized because it is unacknowledged. Knowledge might be more or less observable, more or less complex, and more or less teachable. This problem is central to the discipline of knowledge management.

Knowledge management has been called a management fashion that a lot of organizations adopted throughout the 1990s in order to appear up-to-date (Carter et al., 2008, p. 65). Knowledge management has also been criticized for its alleged ambition, to "suck the vitality from the individual body and soul in order to enhance the vitality of the corporate body" (Carter et al., 2008, p. 64). The academic literature in this area, however, is multifaceted, engaging with philosophy, sociological theory, communication theory, organization theory, technology, and much more (Schwartz, 2006). As such, the knowledge management literature offers more than just utilitarian perspectives.

Let us begin, however, with some basic premises established by the knowledge management proponents. Much of the organizational and managerial literature in this field has revolved around Ikujiro Nonaka's model of knowledge dynamics, which is meant to account for the relationship between tacit and explicit knowledge. Nonaka (1994) developed the model in order to examine what he considered a black box: the information-processing organization. Nonaka sought to pinpoint how knowledge is *created within the organization* – not just how it is *processed by the organization*. At the heart of the model is the concept of 'conversion': "For tacit knowledge to be communicated and shared within the organization, it has to be converted into words or numbers that anyone can understand. It is precisely during this time this conversion takes place – from tacit to explicit, and as we shall see, back again into tacit – that organizational knowledge is created" (Nonaka & Takeuchi, 1995, p. 9).

The model has two dimensions, where the first is related to the conversion of knowledge between the tacit and the explicit levels, and the second is related to the conversion of knowledge from individuals to the organization. The goal of organizational designers is to establish settings where these conversions can take place, for instance by considering how organizational structures allow for knowledge creation. To take a few examples, staff meetings, lunch rooms, and enterprise social media are settings where the issues that employees may be grappling with in their separate offices can be brought out into the open and perhaps benefit colleagues' work. Middle managers play an important role in synthesizing "the tacit knowledge of both frontline employees and top management, mak[ing] it explicit, and incorporat[ing] it into new technologies and products" (Nonaka, 1994,

p. 32). For example, top managers in a private or public organization often have access to customer/client satisfaction surveys. But if approval ratings suddenly change drastically, employees are the ones who are able to account for this phenomenon, as they have experience interacting with customers/clients. Their accounts can be synthesized and made sense of by middle managers. As we can see, this perspective gives more attention to the human dimension than to technology, and Nonaka also calls his model a contribution to a "truly humanistic" knowledge society (Nonaka, 1994, p. 34).

The coupling of tacit and explicit knowledge and the idea of systematically working with organizational knowledge creation has led to the establishment of a large complex of methods, diagrams, and technologies to support the conversion processes described by Nonaka. The aim of knowledge management has been described as providing a comprehensive set of methods for capturing personal and organizational knowledge (Sparrow, 1998). Strategies for administering organizations' knowledge through systematic processes of producing and sharing knowledge (Davenport & Prusak, 2000) are supposed to turn 'knowledge' into 'knowledge-assets' (King, 2009, p. 4). Knowledge-assets are defined as documents such as patents and manuals, electronically stored information, employees' knowledge about best practice, knowledge held by specialized teams, or knowledge embedded in products and services.

Some organizations have appointed knowledge management functions, the purpose of which is to develop the methodologies and systems that can support processes of "knowledge acquisition, creation, refinement, storage, transfer, sharing, and utilization" (King, 2009, p. 4). These elements of knowledge management are often depicted in process diagrams (see for example King, 2009, p. 9), which account for the various steps related to knowledge management. The systematic treatment of knowledge is supposed to help organizations increase performance. If organizations pursue this agenda, a lot of work is required beyond the organization's core task. Knowledge management can thus be seen as a costly practice, often involving the use of consultants and establishment of more support functions in organizations. The same issue of cost presents itself when organizations engage in developing or investing in knowledge management systems, a process we will describe in the following section.

Knowledge management systems

An entire industry has emerged offering knowledge management systems (KMS). Some of the systems are more conceptual, whereas others are based on digital technologies. It has been pointed out how conceptual distinctions from the literature can help us order the functions of the digital systems (Becerra-Fernandez & Sabherwal, 2006). For instance, *Knowledge Discovery Systems* can be databases, data mining, repositories of information, web portals, videoconferencing, email, and so on. *Knowledge Capture Systems* can be expert systems, chat groups, lessons-learned databases, AI-based knowledge acquisition, and computer-based simulations. *Knowledge Sharing Systems* can be team collaboration tools, databases, and expertise locator systems. Finally, *Knowledge Application Systems* can be troubleshooting systems, decision support systems, enterprise resource planning systems, and management information systems (Becerra-Fernandez & Sabherwal, 2006).

In practice, what a KMS might be is still an open question. For instance, a marketing software developer, Hubspot, uses the concept in relation to customer service. They suggest that a KMS can be composed of a range of software solutions that allows an organization's knowledge to be accessed by customers. For instance, an organization's webpage might include items such as FAQ content, forum or community features, 'how to' articles and tutorials, education and training programs, case studies, webinars, and so on (Hubspot, 2017). Sharing knowledge through such web resources is supposed to create value for customers and thereby the business. It also constitutes a potential saving of resources because the need for human interaction in support functions decreases. However, even if knowledge is made accessible on webpages in this way, technology alone does not solve the problem of making knowledge accessible and valuable. People are still central to the production of relevant content, and people are still needed to conduct ongoing assessments and suggest improvements to the KMS.

From early on, the knowledge management literature articulated the concern that an overemphasis on technological systems could lead to an underestimation of human knowledge. Robert Cole expressed it in this way: "While informal mechanisms for the effective conversion of information into knowledge may limit wide dissemination, formal procedures packaged in powerful information technologies often inhibit learning" (Cole, 1998, p. 19). Cole's point is that only humans can ensure that knowledge is embedded in organizational routines. Such understandings are echoed in more recent literature: "Web-based technologies including Web 2.0 and Web 3.0, AI, expert systems, analytics, and collaborative technologies continue to support and transform the field of KM. However, these technologies would not be effective without the day-to-day social aspects of organizations such as 'water-cooler conversations', brainstorming retreats, and communities of practice" (Becerra-Fernandez & Sabherwal, 2015, p. xiii).

Other social aspects have to do with people's use of knowledge sharing systems. For instance, employees might use the systems for purposes other than simply storing and making information available to others. Leonardi and Treem (2012) found that some employees used them as a stage where they could position themselves as experts in relation to their colleagues. Other research has shown that elements such as recognition, monetary rewards, and management support have a positive impact on people's use of web-based KMS such as enterprise social media (Razmerita et al., 2016). It is characteristic of the field of knowledge management that even when scholars explore the potential of digital technologies, they place people in the center. Consider, for instance, the statement that "managing knowledge starts with stressing the importance of people, their work practices, and their work culture, before deciding whether or how technology should be brought into the picture" (Holtshouse, 1998, p. v).

This point was illustrated by a manager in a large public organization, where all texts produced by employees were immediately made available to the rest of the organization, even if they were incomplete and unedited. This means that when embarking on a new project, it is always possible to read the documents produced by colleagues in the area, or that if a colleague is knowledgeable in a field, it is possible to follow the work of this person. As the manager explained, this requires that people refrain from perfectionism and individualism and instead accept being part of a work culture where unfinished work is shared and knowledge is a shared resource.[3]

Knowledge and learning processes are considered quite people-intensive and less technology-intensive (King, 2009). This relationship is reversed when we look into how data and datafication are conceptualized in both theory and practice. In the datafication paradigm, the human-centeredness is replaced with a tendency to neglect the human element in organizational decision-making and value-creation.

Datafication

Most knowledge-based views of the firm seem to operate with either an implicit or explicit hierarchy such that knowledge is higher than information, which is in turn higher than raw data. Hence, *data* is described simply as raw numbers and facts, *information* is understood as processed data; finally, *knowledge* operates at a higher level, as information made actionable (Alavi & Leidner, 1999) (Figure 7.1).

In a sense, this knowledge hierarchy is reversed when we talk about datafication. In this section, we will discuss how data has been elevated to an abstract powerful 'force', which is assumed to 'drive' organizational decisions. When consultants, managers, and scholars advocate data analytics, data-driven decision-making, and uses of big data, we should examine what happens to knowledge, the individual, and the organization. When data is elevated to a force that drives decisions, how do conceptions of knowledge change? What is the role of individual experience in making knowledgeable decisions? And how do organizations manage their data or use algorithmic techniques in everyday practice?

Data analytics has been part of organizational decision-making for decades. Organizations and businesses have practically always collected and analyzed data on customers, business processes, or market economics as

---

[3] This interview with a manager in Local Government Denmark, the Danish association of municipal administrations, is from Plesner and Justesen's ongoing research on the digitalization of public sector organizations.

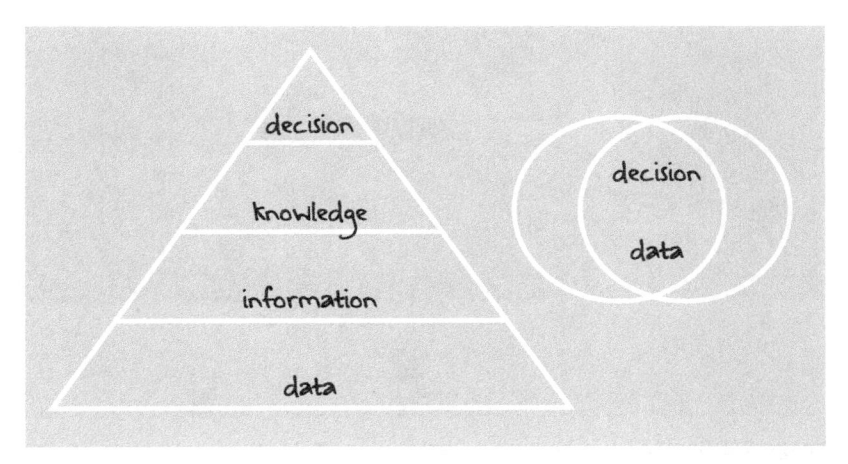

Figure 7.1 The relationship between knowledge and data. As the pyramid illustrates, data has traditionally been regarded as occupying the lowest level in the knowledge hierarchy. In a datafication paradigm, data acquires another status. Here, data practically 'makes' decisions.

a basis for their strategic planning processes. A consultancy business like PricewaterhouseCoopers promises its clients to help them "use data to find the potential to tighten up processes, reduce errors, and cut external spending. Use analytics to highlight patterns that show wasteful or inefficient processes" (PWC, 2017, n.p.).

But when we talk about datafication, this is to suggest that data acquires a new status in practice as well as the scholarly literature, and that data gives rise to new organizational practices (Image 7.2). For instance, the mounting concern with data leads to discussions about what kinds of skills are needed in organizations to create value or stay competitive. In connection with that, data analysts, data scientists, and big data analysts are in increasingly high demand in the labor market. They are trained in, for instance, programming, mathematics, statistics, modeling, or computer science. They either do more basic work, like examining large data sets to identify trends, or advanced work, like developing predictive models. According to the certification training provider Simplilearn, such employees are "the new rock stars" of the era of big data and machine learning (Bhargav, 2018, n.p.). From a scholarly perspective, it has been noted that data scientists have succeeded in establishing themselves as a novel type of expert, gaining public recognition even if they are "data nerds [who] define arcane expertise as theirs" (Brandt, 2016, p. 1), working with quasi-scientific methods. The various professionals working with data have become so sought after because their skills are considered to be crucial for companies eager to practice data-driven decision-making, a buzzword in management for a long time.

Some researchers claim that in the US, for instance, data-driven decision-making has tripled between 2005 and 2010, and seems to have increased productivity (Brynjolfsson & McElheran, 2016). Data-driven

Image 7.2 Baseball has become a datafied sport. Most major league baseball teams employ data specialists who generate and interpret facts and statistics in order to evaluate players and devise playing strategies. The 2011 film *Moneyball*, starring Brad Pitt as the manager of Oakland Athletics, is about this evolution toward quantifying performance.

Source: Peepo, iStock.

decision-making is constructed as directly opposed to the experience-based knowledge discussed in the first part of this chapter: "the companies that had the data they needed and used it to make decisions (instead of relying more on intuition and expertise) had the highest productivity and profitability. Specifically, the most data-driven companies had 4% higher productivity and 6 percent higher profits than the average in our sample, all else being equal" (Mcafee & Brynjolfsson, 2011, n.p.).

The closer we look at the recommendations regarding data-driven decision-making, the more we observe a rejection of 'gut feeling', 'intuition', and 'experience', and the more stories we hear of terrible business decisions made on the basis of such subjective factors. In short, whereas knowledge management valued human experience, the human factor plays a much smaller role in the datafication paradigm.

Big data

Not only 'data', but also 'big data' has become an increasingly prominent topic of debate in many organizations and hence in the scholarly literature. Increased computational power and new possibilities to generate data have led to the storing of great amounts of both structured data (e.g. databases) and unstructured data (e.g. metadata or data generated from social media platforms). This data can be generated either within an organization or elsewhere and then purchased and turned into useful knowledge by the organization. Of course, much of this data just 'sits there', unused, leading management consultants, researchers, and internet evangelists to keep shouting to top managers that they need to leverage big data to keep up with the 'digital disruption'. Recall that we discussed the concept of disruption in Chapter 1, arguing that it is part of a futurist agenda urging organizations to change. When customers and competitors move, a truism in strategy thinking is that an organization will be left behind if it does not change accordingly. This is also the case in relation to big data analysis. Here, some of the recommended strategies to optimize performance on the basis of big data are *predictions, monitoring*, and the *simple display* of complex data. Let us look into a few examples.

Big data analyses in practice: Predictions, monitoring, displays of data

With regard to predictions, retail companies can base their orders on analyses of social media data, web search trends, and weather forecasts. With predictions based on such data, they can ensure that they are carrying the right products in the right amounts. Similarly, a coffee chain like Starbucks can make predictions regarding the profitability of a proposed store location based on traffic data, area demographic, and customer behavior (O'Neill, 2016). In both the retail store and the Starbucks cases, estimates of future needs or future profitability are not based on experience or gut feeling; they are based on predictions using data.

In a similar manner, big data is said to increase efficiency when organizations use it to monitor and adjust operations. Sensors and GPS data can be used to track goods or delivery vehicles and optimize delivery routes by integrating live traffic data. Machines, vehicles, and tools can be made 'smart' (i.e. connected, data-enabled, and constantly reporting their status to each other) so that manufacturing companies can gain real-time visibility into their operations and identify new ways to increase efficiency. These sound like large and expensive projects, but small companies can also develop 'small ideas' based on big data (Donnelly & Simmons, 2013). For instance, one small company uses data from inbuilt sensors to dynamically adjust maintenance cycles depending on how much a customer uses their product. In such ways, small companies can tweak their business models and offer new value propositions. Recall the ideals of organizational learning above – the use of sensors and visualization of workflows are additional methods used to identify areas for continuous improvement.

Both private and public organizations are able to deliver services in new ways by automating the collection of big data and presenting relevant information. Amazon is a well-known example of a company that uses algorithms to collect data and personalize advertisements according to customer histories. Amazon also generates an automated package of all pertinent information whenever a customer calls Amazon's customer service, based on their phone number; this algorithm supposedly allows for a smoother and more efficient conversation between the customer and the Amazon service representative. The same kind of client information package has been launched in some public service organizations, where citizens enter their personal identification number when they call to speak with a caseworker. A robot then opens relevant systems, enabling the caseworker to immediately view all relevant information about the caller.

Data based on sensors, monitoring of workflows, and automated case-handling all come with the added potential that managers can monitor employees in new ways (Plesner & Justesen, 2018). Data can be made visual in quite simple ways by, for instance, depicting a standard workflow. When employees stray away from 'the main road' by spending more time than normally in a given system, or when they open a program that is normally not used to handle a given case, this is depicted visually as detours. The visualizations make it much easier for managers to direct attention to deviations and to save time by not having to monitor work that proceeds according to the established standards of the organization. In principle, this system entails that automated treatment of data 'decides' what managers should spend their time on, and that individual judgment in this domain is needed less. Organizational research has shown, however, that the enhanced visibility afforded by digital technologies in the workplace also creates a number of new managerial tasks (Plesner & Justesen, 2018).

Big data and the question of knowledge

In organizational studies, scholars have pointed out that a by-product of the current hype about big data is an idea that big data analyses somehow produce *truths*. In practice, however, the data does not speak for itself, so we need to gain a better understanding of how organizations deal with data (Hansen & Flyverbom, 2015). Popular accounts of big data tend to portray knowledge production and truth as completely detached from human modes of reasoning. For instance, the former editor-in-chief of *Wired* magazine, Chris Anderson, argues that big data obliterates theoretical models, hypothesis testing, and traditional, more laborious, and imprecise modes of producing new knowledge:

> This is a world where massive amounts of data and applied mathematics replace every other tool that might be brought to bear. Out with every theory of human behavior, from linguistics to sociology. Forget taxonomy, ontology, and psychology. Who knows why people do what they do? The point is they do it, and we can track and measure it with unprecedented fidelity. With enough data, the numbers speak for themselves. (Anderson, 2008, quoted in Hansen & Flyverbom, 2015, p. 883)

Reflecting further on the type of knowledge produced in big data analyses, Hansen and Flyverbom point out that methods of profiling actors on the basis of digital traces and data analytics encourage us to view knowledge and knowledge production in a new light:

> A profile is created by the selection of relevant information. It does not deliver proof of causality or any conclusive reasoning, but mathematical correlations indicative of expected behaviour, by focusing either on patterns or anomalies in the data. The kind of knowledge produced by profiling is different from traditional scientific knowledge that starts out with hypotheses to be tested in search of causes or reasons. (Hansen & Flyverbom, 2015, p. 884)

Recall our discussion of technological determinism in Chapter 3. In contrast to the knowledge management literature discussed above, the techno-determinist literature on big data places agency with the algorithms and the large data sets they process. Big data proponents argue that a new type of knowledge is created, a knowledge based on statistical correlations. This kind of knowledge, they argue, is qualitatively different from the kind of knowledge produced by humans and human reasoning. In practice, we see how organizations attribute knowledge practices normally attributed to humans to algorithms or AI on the basis of large data sets. Consider, for instance, how the tax preparation company H&R Block describes the abilities of Watson, the IBM product we described above, to work with massive data sets:

> H&R Block tax professionals and IBM development teams are training Watson on the language of taxes. They're first applying the technology to the thousands of tax-related questions and topics discussed with an H&R Block Tax Pro during the return filing process. During the tax preparation interview with the tax pros, Watson can *understand* context, *interpret* intent and *draw connections* between a client's responses to suggest credits and deductions that may be available. (IBM, 2017, n.p., italics added)

Notably, 'the tax professionals' are portrayed as playing a major role in the training of Watson, but Watson is then endowed with the capacity to "understand", "interpret", and "draw connections".

The big data hype has generated two very different positions in relation to knowledge: one optimistic and one pessimistic. The optimistic position argues that we can engage with data in empowering ways. For instance, we can generate knowledge about our own bodies and lives that allows us to make informed decisions. This access to and visibility of information offers a democratizing increase in available knowledge. The more pessimistic position holds that the massive generation of data and the development of AI delegates decision-making power to technology. If the collection, calculation, sorting, judgment, and decisions about data are made by algorithms, humans can take it or leave it in the end. Sun-ha Hong has referred to this as 'interpassivity' (Hong, 2015) – a play on the term 'interactivity', which we will discuss in the following chapter on Communication and interactivity. 'Interpassivity' is meant to capture how knowledge and truth are produced in ways that are beyond the scope of our knowledge and understanding – and how humans simply accept this impenetrability, leaving agency to 'something out there'. This view is in line with the critique voiced by mathematician Cathy O'Neill (2017), which we discussed in Chapter 1. O'Neill's argument emphasizes that the problem is not the amount of data but how the data enters into flawed mathematical models and generates invalid results with widespread effects when 'decision-making power' is handed over to digital technologies.

Big data as an organizational phenomenon

The existence of and access to data might be an important part of knowledge creation, but these are not enough. Some kind of critical analysis is also required in order to develop valuable knowledge (Castelfranchi, 2007). This critical analysis is carried out by humans. If big data requires analytical operations, it follows that a first step for organizations is to consider how relevant information and communication can be digitized, how new means of generating data through sensors and digital traces can be utilized, and how algorithms to generate knowledge can be created. But it can also be argued that for organizations, the most important aspect of big data is not the amount of data available or the amount of data that can be generated, but how to develop new ways of thinking about information in order to construct truly useful big data projects and practices. Utilizing big data thus poses entirely different challenges than simply digitizing and thus potentially rationalizing analogue material and processes. The issue is how to construct ways of analyzing and using data. This is the topic of organizational studies of big data.

It has been proposed that to open up big data projects analytically, it is useful to examine four distinct analytical moments in the process of turning real-time data into actionable knowledge (Flyverbom & Madsen, 2015):

1. The production of data (where human conduct and movements of objects are translated into quantitative and binary streams of data that can be stored and processed)

2. The structuring of data (where databases, classification systems, and metadata are chosen)

3. The distribution of data (where data projects are connected to organizational needs)

4. The visualization of data (where data is visualized to give insight into the aspect of the world one wishes to understand).

The point of such an analytical framework is that data is nothing in itself; it needs to be handled and anchored. This again means that data cannot 'drive' organizations (as the expression 'data-driven management' assumes). These handling and anchoring tasks can only be conducted by humans.

Big data and non-transparency in organizations: The case of police work

Let us take the police as an example of an organization that needs to work with large amounts of data of various types. The police also need to translate knowledge into data, as this data can be useful in large-scale data analysis. Much police work has been based on the policeman's judgment and experience, or on knowledge sharing between policemen. Police officers have been trained to make inferences and search for patterns as part of their investigative work. In some areas, their professional practices are still physical or analogue, and documentation is not necessarily standardized. However, it is seen as increasingly pressing to create new information architectures and think about data in new ways. With the aid of computers, it is possible to search for patterns across large data sets. This data search can reveal correlations that can be used to direct investigators' attention.

Such 'predictive policing' is an important trend in police work. The Los Angeles Police Department, for example, uses the PredPol system, which assembles information on previous crimes and produces an interactive map that pinpoints the most crime-prone areas of the city. In New York, the New York Police Department uses their so-called Domain Awareness System to synchronize all the surveillance data of the city with databases of license plates and criminal records to help make their work more predictive (Morozov, 2013). As Evgeny Morozov points out: "Thus, just as Amazon's algorithms make it possible to predict what books you are likely to buy next, similar algorithms might tell the police how often – and where certain crimes might happen again. Ever stolen a bicycle? Then you might also be interested in robbing a grocery store" (Morozov, 2013, p. 184). Like Cathy O'Neill (2017), Morozov is concerned about the effects of delegating power to the automated treatment of large data sets. As he notes:

> Here we run into the perennial problem of algorithms: their presumed objectivity and quite real lack of transparency. We can't examine Amazon's algorithms; they are completely opaque and have not been subject to outside scrutiny. Amazon claims, perhaps correctly, that secrecy allows it to stay competitive. But can the same logic be applied to policing? If no one can examine the algorithms – which is likely to be the case as predictive-policing software will be built by private companies – we won't know what biases and discriminatory practices are built into them. (Morozov, 2013, p. 184)

From an organizational point of view, increasing demands for transparency (which we will discuss in Chapter 9) may push organizations to consider 'opening up' processes of knowledge production that are linked to the use of big data. The non-transparency of algorithmic treatment of big data is well-described in the scholarly literature, and the theme is getting more attention in the public debate.

Knowledge, datafication, and sociomateriality

Organizational scholars have argued that data is not an entity with agency in itself; it acquires agency when it is related to other organizational elements. Data does not 'speak for itself'. Studies of organizational transformations spurred by the datafication trend have showed how both social and technological elements change: organizations not only need to retrain employees, they also find themselves needing wholly new occupational categories and new types of knowledge and expertise. The messiness of data also requires data cleaning and more intuitive work, and organizations need to think about datafication in practical, strategic, and ethical terms. A manager of a public organization described the point of doing big data analysis and data-driven decision-making:

> We use it to think ahead about decision-making. It's simply a combination of using data from our own IT system, using open data, and then we generate some kind of meaning that can form the basis of decision-making. Right now, in relation to data-driven management, we think we've become a bit smarter, *or we can begin to act wiser in the future*. At least in some areas. We can make some more targeted interventions; it becomes possible to work more with prevention, and we can prioritize resources in new ways ... We're very curious about statistics and data for policy committees – imagine if some of this could be used to make some cool decisions for the city? (Manager quoted in Plesner & Justesen, forthcoming)

In practice, studies of the organization showed investments in the datafication paradigm, in the sense that managers chose to hire new data analysts and engage in multiple meetings on data-driven management. Management bought into the idea of data as an entry point to a future in which data would have the status of being able to drive decision-making. At the same time, managers also had very practical concerns about data – its quality, underlying architecture, and the surrounding legal issues (Plesner & Justesen, forthcoming). Again, if we look at data, data-driven management, and big data projects as organizational phenomena, we realize how intertwined technical and non-technical issues are, and how dependent the use of data is on its embeddedness in daily organizational practices.

This may call for considerations of the broader issue of knowledge in organizations. Which types of knowledge are needed to work with data in meaningful ways? Which types of knowledge are produced through the work with data? And since people are needed to work with and around data, which competencies do they need to develop? Such questions also raise collaboration issues. As we mentioned in the first section on Datafication, experts in handling data (statisticians, technologists) are highly valued in various organizational contexts. In most organizations, however, the employees who work with big data projects often make up only a small fraction of the organization, and they need to relate to the other parts of the organization in pragmatic ways.

Strategizing within a datafication paradigm thus needs to consider the practical issues related to reorganization: What are the tasks that need to be solved in the future? Who can solve them? And how should work be organized? These questions are that much more complicated because large data sets and automation of the treatment of data are not simple technical problems – as we have tried to show throughout this chapter. With a sociomaterial, organizational perspective on big data and datafication, we have seen how such technical phenomena are interwoven with people and various organizational elements. We also see that the affordances of new digital technologies have the potential to transform the need for knowledge and expertise within the organization, and that it is necessary to operate with a nuanced approach to the interrelationships between knowledge, data, technologies, and people. They all possess some agency in digital organizing, although this agency is clearly relational (Gulbrandsen et al., 2019).

? QUESTIONS

1. How do organizations understand and/or value knowledge and data, respectively?
2. Do organizations place most resources in knowledge management or data management?
3. How do organizations visualize their work with the production and use of knowledge and data?
4. In what ways are knowledge and data related and in what ways are they disconnected?

References

Alavi, M. & Leidner, D. E. (1999). Knowledge management systems: Issues, challenges, and benefits. *Communications of the Association for Information Systems, 1*(7), 1–37.

Arora, S. (2019). Data science vs. big data vs. data analytics. *Simplilearn*, January 4. Retrieved from https://www.simplilearn.com/data-science-vs-big-data-vs-data-analytics-article

Becerra-Fernandez, I. & Sabherwal, R. (2006). ICT and knowledge management systems. In D. Schwartz (Ed.), *Encyclopedia of knowledge management* (pp. 230–236). Harrisburg, PA: Idea Group Publishing.

Becerra-Fernandez, I. & Sabherwal, R. (2015). *Knowledge management: Systems and processes*. New York: Routledge.

Bell, D. (1974). *The coming of post-industrial society*. New York: Basic Books.

Bhargav, R. (2018). 20 Most popular data science interview questions. *Simplilearn*, October 4. Retrieved from https://www.simplilearn.com/data-science-interview-questions-article

Blackler, F. (1995). Knowledge, knowledge work and organizations: An overview and interpretation. *Organization Studies, 16*(6), 1021–1046.

Brandt, P. S. (2016). *The emergence of the data science profession*. New York: Columbia University.

Brown, J. S. & Duguid, P. (2001). Knowledge and organization: A social-practice perspective. *Organization Science, 12*(2), 198–213.

Brynjolfsson, E. & McElheran, K. (2016). The rapid adoption of data-driven decision-making. *American Economic Review, 106*(5), 133–139.

Carter, C., Clegg, S. & Kornberger, M. (2008). *A very short, fairly interesting and reasonably cheap book about studying strategy*. London: SAGE.

Castelfranchi, C. (2007). Six critical remarks on science and the construction of the knowledge society. *Journal of Science Communication, 6*(4), 1–6.

Cole, R. E. (1998). Introduction. *California Management Review, 40*(3), 15–21.

Davenport, T. H. & Prusak, L. (2000). *Working knowledge: How organizations manage what they know*. Cambridge, MA: Harvard Business School Press.

Donnelly, C. & Simmons, G. (2013). Small businesses need big data, too. *Harvard Business Review*, December 5. Retrieved from https://hbr.org/2013/12/small-businesses-need-big-data-too

Drucker, P. (1957). *Landmarks of tomorrow*. New York: Harper & Row.

Flyverbom, M. & Madsen, A. K. (2015). Sorting data out – Unpacking big data value chains and algorithmic knowledge production. In F. Süssenguth (Ed.), *Die Gesellschaft der Daten. Über die digitale Transformation der sozialen Ordnung* (pp. 140–161). Bielefeld: Transcript Verlag.

Gulbrandsen, I. T., Plesner, U. & Raviola, E. (2019). New media and strategy research: Towards a relational agency approach. *International Journal of Management Reviews*. doi: 10.1111/ijmr.12213

Hansen, H. K. & Flyverbom, M. (2015). The politics of transparency and the calibration of knowledge in the digital age. *Organization, 22*(6), 872–889.

Hasebrook, J. P. & Maurer, H. A. (2004). *Learning support systems for organizational learning*. Singapore: World Scientific Publishers.

Holtshouse, D. K. (1998). Foreword. In U. Borghoff & R. Pareschi (Eds.), *Information technology for knowledge management* (p. v). Berlin: Springer.

Hong, S. (2015). Subjunctive and interpassive "knowing" in the surveillance society. *Media and Communication, 3*(2), 63–76.

Hubspot (2017). *Knowledge management systems: The ultimative guide.* Retrieved from: https://www.hubspot.com/knowledge-management-systems

IBM (2017). *H&R Block's 70,000 tax pros are working with Watson to help clients maximize their refunds.* Retrieved from https://www.ibm.com/watson/stories/taxes.html

King, W. R. (2009). Knowledge management and organizational learning. *Annals of Information Systems*, *4*, 3–13.

Lant, T. K. & Shapira, Z. (2001). *Organizational cognition: Computation and interpretation.* Mahwah, NJ: Lawrence Erlbaum Associates.

Lave, J. & Wenger, E. (1991). *Situated learning: Legitimate peripheral participation.* Cambridge: Cambridge University Press.

Leonardi, P. M. & Treem, J. W. (2012). Knowledge management technology as a stage for strategic self-presentation: Implications for knowledge sharing in organizations. *Information and Organization*, *22*(1), 37–59.

Levitt, B. & March, J. G. (1988). Organizational learning. *Annual Review of Sociology*, *14*(1), 319–338.

Mcafee, A. & Brynjolfsson, E. (2011). What makes a company good at IT? *Wall Street Journal*, April 25. Retrieved from https://www.wsj.com/articles/SB10001424052748704547804576260781324726782

Morozov, E. (2013). *To save everything, click here: Technology, solutionism and the urge to fix problems that don't exist.* London: Allen Lane.

Nonaka, I. (1994). A dynamic theory of organizational knowledge creation. *Organization Science*, *5*(1), 14–37.

Nonaka, I. & Takeuchi, H. (1995). *The knowledge-creating company: How Japanese companies create the dynamics of innovation.* Oxford: Oxford University Press.

O'Neil, C. (2017). *Weapons of math destruction: How big data increases inequality and threatens democracy.* New York: Penguin Books.

O'Neill, E. (2016). 10 companies using big data. *Icas*, September 23. Retrieved from https://www.icas.com/ca-today-news/10-companies-using-big-data

Plesner, U. & Justesen, L. (2018). Merproduktion af målbarhed: Synlighed og nye ledelsesopgaver. *Tidsskrift for Arbejdsliv*, *20*(4), 108–115.

Plesner, U. & Justesen, L. (forthcoming). A multi-sited approach to policy realization and managerial work. In R. Mir & A.-L. Fayard (Eds.), *Routledge companion to business anthropology.* London: Routledge.

Polanyi, M. (1967). *The tacit dimension.* Chicago, IL: University of Chicago Press.

PWC (2017). *Improve business performance – Make informed changes – and make the changes stick.* Retrieved from http://www.pwc.com/gx/en/issues/data-and-analytics/improve-business-performance.html

Razmerita, L., Kirchner, K. & Nielsen, P. (2016). What factors influence knowledge sharing in organizations? A social dilemma perspective of social media communication. *Journal of Knowledge Management*, *20*(6), 1225–1246.

Ross, C. & Swetlitz, I. (2017). IBM pitched Watson as a revolution in cancer care. It's nowhere close. *Stat News*, September 5. Retrieved from https://www.statnews.com/2017/09/05/watson-ibm-cancer/

Schwartz, D. G. (2006). *Encyclopedia of knowledge management.* Harrisburg, PA: Idea Group Publishing.

Sparrow, J. (1998). *Knowledge in organizations: Access to thinking at work.* London: SAGE.

Välimaa, J. & Hoffman, D. (2008). Knowledge society discourse and higher education. *Higher Education*, *56*(3), 265–285.

8 COMMUNICATION AND INTERACTIVITY

> In this chapter, we describe how communication in organizations has evolved as a discipline and how it has been theorized. We introduce interactivity as a strong trend in communication studies and as an empirical phenomenon afforded by digital technologies. We begin by illustrating how traditional organizational communication strategies must take into account an interactive dimension of communication. The chapter then highlights some important traditional subdisciplines and concepts of organizational communication: internal and external communication, strategic communication, and corporate communication. These disciplines and concepts have been challenged by the complexity of communication in practice, and by theoretical developments highlighting the narrative dimension of communication and the material elements that also impact communication. Such developments leave us with an image of organizational communication as increasingly difficult to control and carry out as a one-way process, where the sender formulates a message that reaches a recipient. The chapter proposes that along with our ability to analyze and conduct communication as a planned and tightly managed activity, we should be able to take into account the interactive aspects of communication. Expectations to interactivity and new digital platforms that allow for unprecedented interactivity create new conditions for organizational communication.

Why communication and interactivity?

Both in organizational practice and in theory, communication has been considered an important discipline as well as an important element in constituting organizations. Communication has been seen as a tool for strengthening organizational culture and employee identification, for branding the organization, for managing problematic issues, and for upholding organizational legitimacy generally. Granting communication such a comprehensive role has led many organizations to prioritize the planning and attempts to control communication in and around them. With the spread of digital communication technologies, planning and managing organizational communication has become more challenging. For instance, it has become obvious that social media cannot be used simply as conduits for an organization's message. Social media and other digital communication platforms transform organizational communication into an interactive phenomenon.

The challenge of interactivity to controlled communication: The case of Twitter

Let us take the example of Twitter, which has become an increasingly popular channel for creating awareness of corporate messages and activities. Oana Albu and Michael Etter (2016) studied how organizational members and non-members are active in constituting organizations through communication on Twitter. People can either retweet messages loyally and thus help disseminate them, or they can modify messages by adding critical remarks and thus participate in the destabilization of the reputation of the organization by spreading these to new audiences. In Albu and Etter's study, managers were acutely aware of the change in organizational communication practices brought about by social media. As one of them said: "If you want control, don't turn the computer on. Here you cannot control the messages in a way that traditionally you could" (Albu & Etter, 2016, p. 25). The manager saw Twitter as dangerous terrain and the hashtag as an uncontrollable feature. At the same time, managers realized that they could not avoid social media. As another manager said: "We really focus on how we communicate on social media because people now are on Twitter all the time. But you need to be careful because it is not a one-to-one conversation like we are having now. *It's 'puff'… it can go anywhere*" (Albu & Etter, 2016, p. 20).

Albu and Etter point out that it is the hypertext aspect that makes Twitter a unique kind of communication site, because people can intentionally modify and then redirect a text with a new meaning to a new readership. The new message may be ironic, for instance by retaining elements of the original message, but giving it a critical twist. As such, the hypertext function is central to disrupting our usual assumptions regarding sole authorship and organizational control of its communication. And all of a sudden, management's attempts to carefully curate or maintain a desired image and reputation of the organization through communication becomes not only very complicated, but risky. One of the examples given in the study is that of a fast-food restaurant chain (known by the pseudonym Beta) which launched a campaign with the hashtag #BetaSupply. This hashtag was retweeted 1,400 times, and 76 percent of the retweets contained negative statements such as "I'm just getting coffee, but I feel dirty being here at #BetaSupply" or "#BetaSupply: Liquid chicken nuggets. Who's hungry for Beta?" (Albu & Etter, 2016, pp. 21–22). The organization tried to launch a new hashtag to draw the attention away from the first one. Under the hashtag #MeetOurSuppliers, the tweet said: "He's what you'd call a Beef Snob. Meet Steve, our beef supplier who always delivers on quality #Meet

OurSuppliers", but no one retweeted it. Instead, the negative comments continued under the first hashtag, and the entire episode ended up becoming a story in the mainstream media (Albu & Etter, 2016, pp. 22–23). Albu and Etter consider this case to be an example of how organizations become 'co-produced' through digitally mediated communication in which many different actors participate. Moreover, the case exemplifies how social media has made the old-style, one-way communication nearly impossible. For better or for worse, we live in a world of 'interactivity'.

In this chapter, we discuss the consequences of such interactivity afforded by digital technologies, and the conditions that interactivity creates for organizations and for our understanding of organizational communication. We will argue that some of the traditional communication disciplines – such as strategic communication or issues management – are still important domains to develop, just like traditional communication concepts have an analytical value. However, addressing these areas and using these concepts requires an awareness of the fundamentally interactive character of communication. This new condition becomes much more obvious when we look into the communication practices afforded by digital technologies.

Communication as a theme in organization studies

As we saw in Chapter 4 in our discussion on structures, early organizational scholars realized the importance of ensuring efficient lines of communication (e.g. Barnard, 1938; Chandler, 1962). However, the way we might understand communication in organizations as a phenomenon or a particular discipline was not a topic of sustained interest. With inspiration from Communication Studies, there emerged a new field called 'Organizational Communication'. Initially, communication was mostly understood as simple message transmission, and the field had the same focus on organizational efficiency as did early organization studies. Thus, scholars were concerned with measuring the relationship between communication effectiveness and organizational effectiveness, and they were able to argue that the relationship was strong (Morley et al., 2002). Otherwise, organizational communication was often considered a part of Communication Studies rather than Organization Studies, and within organizations, communication was to some degree an isolated activity carried out by communication professionals.

The institutionalization of organizational communication *studies* evolved simultaneously with an upsurge in the organizational communication *profession* (see for example DeWine, 2001). Both private and public organizations created or expanded their communication departments, gave priority to crafting communication strategies, employed cadres of communication professionals, and began to provide media training to managers and even some lower-level key employees. This prioritization of communication can be linked to increasing demands that organizations appear as legitimate actors in society, and more recently, to online activism that forces organizations to open up and become more accountable (Sifry, 2011). We will discuss this development in more detail in Chapter 9 on legitimacy and transparency. Since organizations need to maintain their license to operate by being perceived as legitimate, organizations have designed communication strategies to either frame issues in favorable ways, brand the organization, or secure employee alignment. The explosion in organizational communication activities can also be linked to the information society, as the increasing amount of available information practically begs for someone to select, frame, and craft this information into meaningful messages. Here, we will not discuss to what degree communication is always the correct tool to apply, but simply note that communication has become increasingly popular as a solution to a range of problems.

Along with the increased amount of resources spent on the professionalization of communication in organizations, it has also become commonplace to demand strong communication skills of employees in many parts of the organization. As Charles Conrad and Marshall Poole (2005, p. 4) note: "Of course, employees cannot function effectively unless they possess the technical skills that their positions require. But more and more it appears that being able to recognize, diagnose, and solve communication-related problems is vital to the success of people in even the most technical occupations".

It has apparently become common sense that communication is an integral part of enhancing efficiency, as pointed out by the early organizational communication scholars. At the same time, we can also link the increasing priority placed on communication to the trend, pointed out earlier, toward 'flatter' organizational structures and decentralized problem-solving. If managers are no longer solely responsible for making the right information available and ensuring that complex tasks are solved in the best way, this 'freedom' creates new demands on employees – including new expectations of communication skills on all levels of organizations. Consider, for instance, the advertisement for a job as an IT specialist in the US Justice Department. Among the basic qualifications for an applicant, one of the bullet points lists: "Oral Communication – Expresses information to individuals or groups effectively, taking into account the audience and nature of the information; makes clear and convincing presentations; listens to others, attends to nonverbal cues, and responds appropriately" (USAjobs.gov, 2016, n.p.).

Such demands in a job advertisement can be linked to the idea that when employees communicate competently, this makes the organization more efficient. The demands also illustrate a heightened concern with the need for individuals central to the organization's operations to interact productively with one another. If we reflect on the search for employees with the qualifications described in the job advertisement, it might seem self-evident that organizations will be less successful if their employees do *not* express information effectively, do *not* take into account the nature of the audience, do *not* make clear and convincing presentations, do *not* listen to others, and do *not* attend to non-verbal cues. What is not self-evident, however, is precisely how organizations choose to ensure that these communication tasks will be handled competently. Often, they turn to traditional communication concepts and traditional communication disciplines, which are not only sets of tools, but also topics of scholarly concern. In the following sections on Internal and external communication, Strategic communication, and Corporate communication, we discuss three such disciplines.

Internal and external communication

Traditional organizational communication would have at least two important subdisciplines: internal communication and external communication. Work with internal communication might consist of producing informational material based on top-management decisions and activities or creating outlets for communication among employees. A typical example are the weekly or monthly newsletters circulated to all employees by email, informing them about management decisions, special events, the menu in the cafeteria, and other important issues. Work with external communication would consist of crafting press releases or in marketing campaigns. In both instances, practical work with communication demands rather technical skills in writing texts or oral presentations, the work of structuring, wording, presenting, and 'getting your message across'. At the core of this 'technical discipline' are the fundamental elements of the communication process: Who says What through Which Channel to Whom with What Effect (Lasswell, 1948). If we translate this basic formula into the work of communication professionals in organizations, we can see that they work with these basic elements at an organizational level: They attempt to create particular images of the sender of communication (be it the organization or the management). They craft and test messages. They select or develop appropriate media for communication. They analyze target groups and work to make sure that the message and the communication channel are appropriate in relation to that target group. And, finally, they test whether or how communication has an effect on the target group.

It should be mentioned here that the internal/external distinction is not unproblematic. It has both been challenged in communication theory (e.g. Cheney & Christensen, 2001) and by empirical phenomena such as US president Trump's tweets about classified information (LoBianco & Raju, 2017), which disregards any boundary between what is otherwise considered internal/external. It remains that communication activities directed at employees and other stakeholders has become an important subdiscipline in the more managerial communication literature. It is connected to the broad practical disciplines, strategic communication and corporate communication.

Strategic communication

Strategic communication can be seen as "the purposeful use of communication by an organization to fulfill its mission" (Hallahan et al., 2007, p. 3). It thus links organizational strategy to communication, suggesting that it is necessary to establish overall goals for what communication is meant to accomplish for the organization, just like it is useful to develop plans for how to approach communication strategically.

As an academic discipline, strategic communication is relatively young. The field has been dominated by a prescriptive approach, which sees organizational communication as carrying out situational analyses, making decisions, and then implementing the communication strategy. In this line of thinking, communication is a support function for top-management and is the responsibility of the communication executives. From around the year 2000, there emerged new types of academic perspectives on strategic communication, examining how it is done in practice, and insisting that even strategic communication emerges in everyday practice and is intertwined with organizational learning and unpredictable events (Frandsen & Johansen, 2014). This development has led to the idea that instead of just 'having' a communication strategy, it might be more important to think strategically about communication at various levels and moments in the organization. Gulbrandsen and Just (2016) propose to 'strategize communication', that is to emphasize the understanding of, management of, and participation in strategizing processes. They talk about this as *communication strategy in the making*. They are among those scholars who have recently criticized communication strategies for being too static and insensitive to real-life problems.

In organizations, we sometimes observe the same skepticism. A head of the communication department in a large organization explained that he does not see the value in communication strategies. He explained that this is not because organizations should be non-strategic about their communication, but because focus needs to be placed on the corporate strategy, which should lay the foundation for daily communication activities. In this way, communication can be approached as a more dynamic process, where employees address communication tasks on the basis of the corporate strategy.[1] This approach, linking communication to the corporate strategy, demands considerable responsibility on the part of employees – quite in line with the point we discussed in relation to the less hierarchical and more network-based organizations in Chapter 4. In such strategizing processes, it is common to have morning meetings, weekly meetings, or workshops to discuss issues and problem-solving in order to ensure that employees are equipped to make their own decisions. In this particular media organization, an employee explained how difficult it is to work professionally with communication if there are no guidelines as to how it should be carried

---

1 This example is borrowed from some of our students' work. We do not credit them here because their study is confidential.

out. In the absence of such guidelines, it is left to the employee's common sense to communicate appropriately on behalf of their organization. This freedom increases the probability of conveying an overly complex image of the organization to its stakeholders. It is yet another indication that tightly controlled communication is hard to achieve in practice.

Corporate communication

In opposition to this way of thinking and acting in relation to organizational communication, the discipline of corporate communication is based on the idea that even complex organizations ought to speak with a single voice. The word 'corporate' comes from 'corpus', which means body. This metaphor of the corporate body is central to the discipline of corporate communication because it conveys the ideas that 'one hand should know what the other hand is doing', or 'the mouth should not say something while the body demonstrates the opposite'. Corporate communication has been defined as: "The set of activities involved in managing and orchestrating all internal and external communications aimed at creating favorable starting points with stakeholders on which the company depends" (Riel & Fombrun, 2007, p. 25).

The ideal is that an alignment is established between the messages of top management, various departments, and individual employees so that the organization does not appear fragmented or even untrustworthy, as would be the case if it sent mixed messages to its external environment. One example could be if a company has developed a high-end brand through marketing activities – for example through the use of storytelling and images – and the employees who answer the phone give incorrect advice or bad service when customers call. These would be seen as conflicting accounts, and the latter would be seen as undermining the former.

Communicating strategically – in an 'integrated' fashion – is also seen as increasingly important for business to business communication in a changed media landscape. Formerly, businesses communicated to their customers using a combination of direct mail, paid media campaigns, media stories planted through public relations channels, and industry fairs. Today, organizations have websites, social media profiles, and other media, which they own and control. As marketing blogger Tim Asimos notes, these new media possibilities have "marked a seismic shift in marketing communications", because now communication is not about 'outspending' competitors, but 'outsmarting' them by creating remarkable content that can be shared. As he writes, this can give just as much marketing power to small- and medium-sized companies as large ones. Even in business to business communication, it is relevant to create a convincing online presence, because businesses also do online research on their partners. With the opportunities afforded by new media, however, also comes the risk of creating a fragmented image of the organization, so that all the communication efforts should be "operating from the same overarching communications strategy" (Asimos, 2017, n.p.) – in line with the corporate communication ideal.

The idea behind corporate or integrated communication is that the organization as a whole risks being considered untrustworthy and hence illegitimate when it does not speak as one body – or with one voice. Recently, some organizational scholars have claimed that this more fragmented communication need not pose a problem. They have criticized the ideal of speaking with one voice as being rather naïve, as buying into a consistency trend that they find problematic (Christensen et al., 2008). The critics have argued that communication is inherently multivocal – or 'polyphonic' – and should not necessarily be reduced to a question of sending uniform messages. Rather, organizations should learn to take advantage of difference and to accept the fact that there will always be gaps between, for instance, the articulated vision of the organization and the myriad of activities and utterances that otherwise define the organization (Christensen et al., 2013).

The normality of such gaps becomes obvious in a study of the communication in and around a large pharmaceutical company, Novo Nordisk. Here, Gulbrandsen and Just (2013) identify three dominant narratives: (1) a socially responsible organization, engaged in society at large, (2) an organization concerned primarily with fulfilling its own objectives – profit maximization, and (3) a great and employee-centered workplace. The authors' idea is that Novo Nordisk has to acknowledge all these voices and maintain a polyphonic communication with their stakeholders in order to appear transparent and responsible. The analysis of Novo Nordisk is based primarily on online communication, and to be sure, many of the challenges to the corporate communication ideal are tied to the role of digital infrastructures such as social media, online communities, chat with employees, and other digital interfaces that organizations have with their stakeholders.

New conceptions of communication

Our discussion of traditional organizational communication disciplines has already indicated that it has become both empirically and theoretically untenable to conceive of communication as a one-way process that organizations can somehow control. Communication studies have become increasingly occupied with analyzing and theorizing communication as an interactive process of meaning-creation, much more dialogic and polyphonic in nature than traditional transmission perspectives would have it. The academic discipline of organizational communication has also grown a lot broader. For instance, it has studied how communication reflects or reproduces power relations, gender relations, or conflicts. From the 1960s and 1970s onwards, the field also became more attuned to how its own theoretical development affected how we think about communication and how organizational communication impacts organizational life. As Steve May and Dennis Mumby (2005, p. 8) write in their introduction to critical organizational communication studies: "Developments in postpositivist, critical, postmodern, and feminist research, among other developments, have greatly enhanced our insight into the communicative dimensions of organizing. For example, recent feminist research has moved us beyond treating gender as a variable, toward recognizing the fundamentally gendered character of the organizing process".

This development has led scholars to distinguish between a communication discipline occupied with non-reflective 'How-To-Do-Issues' and with reflective-critical issues, respectively (Klikauer, 2007, p. 11). This might be a useful distinction, conceptually. However, we do not wish to suggest that it is irrelevant to discuss the traditional concepts that highlight the elements of communication, or irrelevant to be technically adept at communicating important issues in organizations. What is important here is the growing understanding that organizational communication is much more than this. We will highlight two traditions within organization studies that have grappled with understanding communication as a broader organizational phenomenon: the narrative-discursive approaches and the material approaches.

Narrative approaches

The idea that communication is not just an add-on to other organizational activities builds on the assumption that language is not a neutral medium for conveying messages, but constitutive of social reality (Berger & Luckman, 1966), and, in this line of thinking, constitutive of organizations. When we choose to use certain terms, when we categorize, when we evaluate, and when we justify decisions, this creates a particular reality within which we act. To take a few examples, talking about employees as members of a 'family' conveys the expectation that employees are committed, loyal, and caring in relation to the organization as a whole (Costas, 2012). Or, adding 'junior' or 'senior' to job titles both supports the hierarchical structures of an organization and justifies different privileges and the size of paychecks. Or, when an 'employee of the month' is selected and put on display, this illustrates the values guiding the organization, and the precise way decisions and actions are justified adds to the process of defining what kind of organization we are dealing with. These types of observations are grounded in discourse analysis, which has become increasingly popular as an approach to understanding organizations (Cederström & Spicer, 2014; Fairhurst & Putnam, 2004).

Discourse can be defined as patterns of utterances that either reproduce or challenge our social reality. Discourse is thus not *the same* as communication, but communication is discursive because according to discourse theory, language use is constitutive of knowledge, identities, and relations (Fairclough, 1993). Organizational scholars "increasingly assert that organizations are discursive constructions because discourse is the very foundation upon which organizational life is built" (Fairhurst & Putnam, 2004, p. 5). In organization studies, different terms have been used to convey this idea. Karl E. Weick talked about *sensemaking* as a central activity in the construction of the organization and the environment it confronts (Weick, 1995). In other corners of organization studies, it has been argued that the main source of knowledge in the practice of organization is *narrative* (Czarniawska-Joerges, 1997), or that *organizing happens in communication* (Clark et al., 2011). The idea that 'organizing happens in communication' has achieved considerable attention because of its radical claim that there is no organization beyond the communication that creates and sustains it over time. From being a technical discipline carried out in particular parts of the organization, communication is suddenly seen as the very essence of organization rather than just a variable (Putnam & Nicotera, 2009). The so-called 'Communication as Constitutive of Organization' (CCO) scholarship thus focuses on how communication is used by human beings to coordinate actions, create relationships, and maintain organizations.

Material approaches

Although the narrative turn in organization studies has been acknowledged for pointing out the effects and implications of language use for power and identity-formation, the narrative turn has also been criticized for its exclusive focus on the symbolic realm and neglect of the material dimensions of organizing (Carlile et al., 2013). In the context of our discussion about communication, it is essential to view materiality as intertwined with meaning. Meaning-making and communication have always had a material dimension. To give an example, in creative work such as design, arts, or technical innovation, objects or material mock-ups are often a necessary part of the work process. But rather than being purely material, these materials are completely intertwined with the meaning that actors ascribe to them, and with the stories that actors are able to tell about them – "meaning and materiality are constitutively entangled" (Gheradi & Perrotta, 2013, p. 254).

It has almost become a truism that artifacts in organizations have an important symbolic function (Schein, 1985). Consider, for instance, the use of gifts or goodie bags to create bonds with customers, the display of art in corporations as a marker of cultural status, or the use of alternatives to furniture such as gym balls or the like, to encourage creative thinking. However, there is also a range of artifacts or objects that have a very practical impact on organizational life – at the same time as they are either thoroughly symbolic or otherwise connected to communication. For instance, strategy tools (Belmondo & Sargis-Roussel, 2015), performance management tools (Brogaard-Kay, 2015), and technical drawings (Ewenstein & Whyte, 2009) have been theorized as social/material constructions that can highlight organizational practices such as decision-making, performance conversations, and collaboration. They have both a linguistic dimension – being constructed in language and symbols that interact with wider social structures, and a material dimension – often reified in the form of printouts so that people can literally gather around them, point to them, and so on.

Similarly, architecture and the physical workspace have been theorized as communicative because they convey 'organizational meanings' or communicate an organization's brand or goal (Marrewijk & Yanow, 2010).

As a response to the material turn in social science – and organization studies – Organizational Communication scholars have strived to redefine 'organizational communication' as a much broader issue than what has been portrayed by the technical disciplines or the narrative approaches. In its broadest definition, organizational communication is an *"ongoing, situated, and embodied process whereby human and non-human agencies interpenetrate ideation and materiality toward meanings that are tangible and axial to organizational existence and organizing phenomena"* (Ashcraft et al., 2009, p. 34, *emphasis* in original).

To sum up, the narrative and material approaches to communication in organization have shown us that communication is neither epiphenomenal to organization nor is it purely linguistic. These insights lay the foundation for discussing what happens to communication as increasingly important digital infrastructures are established in and around organizations. Such infrastructures afford types of language use and communication patterns that have not necessarily been common or visible in organizations. It seems fair to suggest that in both theory and practice, communication needs to be conceptualized not so much as a matter of transmitting messages but as a form of interactivity; this is especially prominent when we take into account the intimate relationship between communication and digital infrastructures. At the end of the chapter we will further argue that along with new issues arising around communications in organizations, more traditional patterns and practices continue to exist. This is why we need to be able to work analytically with both 'communication' and 'interactivity'.

Interactivity

In our discussion of the development of various disciplines in organizational communication and strategic communication, we observed an evolving awareness of the uncontrollability of communication. We saw attempts to understand the complexity of communication processes when many actors are involved and when organizations and their environments are increasingly complex. This complexity becomes accentuated when we consider how digital technologies affect organizational communication. Ib T. Gulbrandsen and Sine Just (2016, p. 25) describe how we need to see communication differently in the light of new media:

> we can no longer reduce communication to a linear and singular movement of one message from (identifiable individual or group) A to (identifiable individual or group) B. Instead, it is now imperative to fully conceptualize communication as process in a more radical sense: a constantly evolving movement in and

> between many utterances and in and across dispersed networks; a movement in and through which various identities are articulated and deployed using different off- and online props; a movement that plays out as a 'never-ending' present into which we can recover and rearticulate past statements at any moment in time; a movement that waxes and wanes, sometimes buzzing with activity, more quiet at other times, sometimes apparently at rest, but with the constant possibility of re-activation.

Although this description of communication might seem quite radical in its process orientation, the authors do not suggest that new media have necessarily or fundamentally changed the way we live, work, and communicate. Instead, they propose that new media both enhance already existing communicative practices and enable new ones (Gulbrandsen & Just, 2011). We can thus conceptualize communication as a more dialogic and complex practice that has been underway for decades. The way communication takes place on digital platforms and in digital networks reveals how communication is inherently, thoroughly interactive. The term 'interactivity' might therefore be an accurate description of what is particular about communication as an element of 'digital organizing'. But interactivity is not only a property of technology. Instead, it is something that changes our perceptions (Bucy, 2004) of what comprises communication in organizations.

Interactivity in communication theory

As early as the 1960s, reception studies (e.g. Hall, 1963/1980) argued that those who were called 'receivers' of communication were actually actively shaping communication messages rather than just absorbing them. This recognition of the receivers' active role led communication researchers to talk about 'users' rather than 'receivers' of communication. At the same time, the idea of 'target groups' also began to be challenged. If individuals 'used' communication for their own purposes, based on their backgrounds and interests, it was less likely that they could be viewed as homogenous groups of people with the same preferences, as implied by the target group thinking. As the notions of 'receivers' and 'targets' of communication began to appear theoretically untenable, the ideas of 'coproduction', 'interactivity', and 'dialogue' surfaced in communication studies (Phillips, 2011). These concepts point to the collaborative process that brings meaningful communication into being. This process is derived from three basic assumptions: (1) Meaning is co-produced by the parties involved in the communication process. (2) Communication is, inherently dialogical because it always presupposes a partner. (3) There needs to be activity among people to bring about communication, that is, communication presupposes engagement.

With more interaction, we also see a possible empowerment (Barry & Doherty, 2017) of individuals, who are suddenly seen as capable of *acting* (see also Chapter 10). Such ideas have had consequences for how most scholars now think about planned communication. Planned – or strategic – communication has to take into account the fact that communication is simply more interesting when it is relevant to the people it addresses, and that relevance often arises from participating in the communication process, or being part of shaping it directly. Empirically, we can also observe that people have become less willing to just receive information; for instance, traditional media such as flow television and newspapers have a declining 'audience', and the activity of 'users' has become visible when observing social media platforms because much of the engagement with content happens in these collective digital spaces. Such developments have led some scholars to completely abandon the concepts of 'receivers', 'audiences', or 'users' and to instead talk of 'commons-based peer production', 'consumer-generated media', or 'producers'.

Interactivity in media theory

Let us dwell a bit on the concept of 'producers' (which we also discussed in Chapter 5). Recall that 'produsage', coined by Axel Bruns (2008), is a linguistic blend of the words 'production' and 'usage'. The term is not only relevant in discussions of production, but also in relation to changing communication practices. It captures well the phenomenon of interactivity, which points to the fact that it has become increasingly untenable to think in terms of senders (producers of content) and receivers (users of content). Fewer people are passively 'consuming' media content, and more people are actively engaged in producing it. The idea of produsage – and other terms that denote interactivity – came out of a media studies tradition.

Media scholars have traditionally been interested in the relationship between media production, media technologies, media content, and media consumption. Prior to the internet, the production and consumption of mass media and the ideological influence of mass media would be central areas of interest. With the growth of the internet and its interactive applications, media studies has become much more interested in what has been termed the 'participatory culture' afforded by the internet (Jenkins, 2006). Media scholars point out that web 2.0 applications have enabled the production and distribution of uncensored content; anyone can contribute to news production and commentary with video footage, Facebook posts, or comments on news websites; anyone can distribute creative work on YouTube, Instagram, or blogs; anyone can intervene in political debates or engage in special interest groups with like-minded people on various platforms. And this participation is not just a technical possibility; it is increasingly becoming a way of being a member of society. Or at least a member of a subsection of society – since much interaction happens in 'echo-chambers' created by social media algorithms that keep feeding people the type of content they already read and like.

The contribution of media studies to the concept of interactivity (e.g. Kiousis, 2002) is to conceptualize interactivity as part of a cultural

COMMUNICATION AND INTERACTIVITY

transformation that deeply affects societies. Interactivity changes the way individuals and groups interact, and, by extension, alters communication practices and expectations to communication (Image 8.1). This reality penetrates organizations and organizational communication; organizational members and organizations' stakeholders are also nurtured by a participatory culture and have come to expect that communication is an interactive phenomenon.

Organizational communication and interactivity

Organizations still employ communication professionals and still engage in internal and external communication practices, including strategic communication and corporate communication. In addition, it is still the case that communication can be seen as a constitutive element in organizations. But all these communication activities – and communication as such – have obtained a new dimension because of digital technologies. The affordances of digital technologies create unprecedented interactivity and hence alter communication in organizations.

As we saw in our discussion of Twitter at the beginning of this chapter, digital technologies afford a blurring of boundaries between the organization and its environment (Gulbrandsen & Just, 2013; Plesner & Gulbrandsen, 2015). We also saw the challenges faced by organizations when using social media platforms such as Twitter or Facebook for one-way, one-to-many dissemination of messages. The platforms are created to afford interactivity, and their users are part of a participatory culture where interaction – commenting

Image 8.1 Mobile devices allow us to be 'always on', and because many people are always on and practically always connected through social media, organizations have many opportunities to reach their stakeholders using different social media platforms.

Source: ViewApart, iStock.

187

and sharing – is the norm. Of course, Twitter and Facebook are not constructed alike, and their usage is not always marked by interactivity; US president Donald Trump's use of Twitter has proven that it is possible to use this platform to set agendas in very effective ways. However, the difficulty in disseminating one-way, uncontested messages and the interactivity now marking organizational communication pose important questions about controllability. Large organizations that have many interfaces with a large number of stakeholders have probably always had problems creating a unified corporate communication, but these problems were more manageable when conflicting narratives were limited to dinner-table discussions or occasional mass media stories. By contrast, when organizational communication becomes collaborative and visible on online platforms, when contesting accounts are leaked and commented on, it becomes more difficult to handle issues communicatively, and to uphold a unified image of the organization. And in large organizations, when different divisions, offices, or other organizational units have their own online presence or become objects of public discussion, any effort at communicating a message can evolve into a communication situation so complex that it is impossible to manage in any simple way.

Social media do not just pose a challenge to organizations by making their communication less controllable. Because people spend so much time on these platforms, they offer possibilities of reaching a much larger group of stakeholders than other media (Cornelissen, 2014). If organizations are to take advantage of this potential audience, having a presence on these platforms is not enough. They also need to acknowledge that communication is an interactive process. This means that they must be able to invite people into genuine dialogues about ideas, participate and encourage debates, and strive to create the kind of content that is worth sharing and liking. For instance, many companies ask people to share photos and stories about their use of the company's products, municipalities provide online spaces for citizens' complaints and ideas for improvements, and patient organizations create discussion forums where people can share experiences, which can become catalysts to the political work of the organization.

Interactivity and relevance: The case of IKEA

Creating space for interactive communication demands reflections on what could seem relevant, interesting, or entertaining for a large number of individuals, rather than remain sender-oriented, hammering out a message to which people do not react. A means to acknowledge interactivity – the necessity of engaging conversation partners – is by offering services that create value or occasions for action for people, rather than offering 'content to be consumed'. An example here is the furniture retailer IKEA, who was early to offer their customers three-dimensional kitchen drawing programs online. People could use this design software for free, without going to the store, or ever buying a product. This can be considered a communicative practice because it supports IKEA's narrative about offering 'democratic design' (i.e. affordable products), while also serving the purpose of displaying some of their kitchen products and expertise. We discussed the material aspect of communication above. Every element of IKEA is communicative; from the warehouse layout to the packaging of products, all of which signal affordable and 'no-nonsense' products. Other communicative aspects are the free childcare facilities in the store and the in-store cafeteria. They are investments that communicate how IKEA (2019, n.p.) "understand[s] the needs of families with kids", as the company states on its website. Apart from this, IKEA spends considerable resources on social media – they are present on YouTube, Instagram, Pinterest, and Facebook. IKEA's Facebook page is 'liked' by 27 million people, and there is quite a lot of activity there. Besides advertising and recalling defective products, IKEA also tries to use Facebook more dialogically. For instance, they ask their customers questions about which new food items they would like to see on the menu in their customer cafeteria. The company normally responds within a day, so it is obvious that customer feedback is given priority by the organization. Offering online resources for free and maintaining social media dialogues are communication initiatives that demand resources, but given expectations to interactivity, they presumably pay off.

Besides the social media that crisscross organizational boundaries and influence communication between employees and stakeholders outside the organization, multiple digital technologies have been established to specifically support communication *within* organizations. For instance, intranets and various types of knowledge-sharing platforms make up elements of the IT infrastructures we discussed in Chapter 4. They have different functionalities and organizational implications, but if we take Enterprise Social Media as an example, it has been suggested that such media "have significant implications for communication *inside* the workplace, influencing such organizational communication issues as interaction with new hires, knowledge sharing and management, and employees' abilities to form relationships and build social capital" (Leonardi et al., 2013, p. 16). Enterprise Social Media mimic the publicly accessible social media in layout and functionalities, but they operate as non-public communication platforms. As Paul Leonardi and colleagues underscore, they do not function as channels through which communication travels, but as platforms on which social interaction occurs. Anyone in the organization can participate at any time from any place. Leonardi and colleagues identify numerous both positive and troubling aspects of introducing Enterprise Social Media in an organization. For instance, it is positive that they give insights into what others are doing and who they know. In that way they help create "conversational fodder that makes it easy to initiate new connections" (Leonardi et al., 2013, p. 8). But there might be a disadvantage in that "peripheral awareness of others may create illusion that a real social connection exists when it does not" (Leonardi et al., 2013, p. 8). We will not go into any evaluative discussion here, but simply note that organizations that introduce such digital technologies must be aware of how this opens up and broadens out communication and that it supports an interactive organizational communication environment. This awareness has entered the management literature, where managers are advised to adjust their communication style and embrace interactivity. For instance, Srinivas Koushik and colleagues (2009) have put forward this list of suggestions:

1. Adapt to the new world – tap into the collective wisdom of the team
2. Challenge existing modes of communication – become personal and close to audiences
3. Work on being modest – get used to transparency and the feedback this results in
4. Be an early technology adopter – connect with the younger workforce.

These suggestions implicitly refer to many of the topics that we have discussed in this chapter: that top-down, one-way communication is untenable and that employees increasingly expect to participate and be active in framing organizational issues. That this new style of management is called 'Management 2.0' (Koushik et al., 2009) indicates that in the authors' minds, the changes in communication dynamics are linked to the evolution of digital technologies.

New digital technologies and interactivity not only have implications for managers, but also for employees. For instance, internet-connected mobile devices and email apps mean that employees are potentially constantly connected to their workplace (e.g. Orlikowski, 2007; Villadsen, 2017) and that they are expected to reply to all kinds of questions and issues on a 24/7 basis. When coupled with the participatory ethos that comes with digital technologies, connectivity and mobile devices become important elements in the blurring of boundaries between employees' work life and private life (Gant & Kiesler, 2001; Mazmanian et al., 2013), as well as in extending organizational communication far beyond the formal boundaries of the organization (Plesner & Gulbrandsen, 2015). It has become a topic of debate when the seeming luxury of replying to emails from the beach or while sitting in a café turns into a work life that takes over all our waking hours. Some organizations are so worried about their employees' chances of resting and staying healthy that they forbid their employees to send or reply to emails on Sundays or turn off the internal communication channels after a certain hour at night (see also Chapter 11).

So where do all these developments leave us regarding the relationship between communication and interactivity? Some organizational scholars argue that new communication practices have fundamentally changed organizations' relations to their employees, their consumers, and with the public at large: "Social media have altered the everyday communication that occurs both internally and externally, by enabling an interactive and dynamic process to occur between individuals and the organization" (Manuti, 2016, p. 19). Nevertheless, while acknowledging the changes wrought by digital

Table 8.1 Perspectives on communication in organizations and their core concepts.

Perspectives on communication	Core concepts
Communication as tools to ensure organizational efficiency and legitimacy	Internal communication External communication Strategic communication Corporate communication
Organizations as communicative entities	Sensemaking Narratives Communicative constitution Materiality
Communication as interactive	Dialogue Produsage Participatory culture Interactivity

communications, these scholars still believe that, for instance, social media strategies should consist of the same elements as a traditional communication strategy: the strong focus on target groups, goals, channels, and so on (Manuti, 2016). We believe that the recognition of digitally afforded interactivity ought to go hand in hand with a recognition of the *fundamentally* interactive character of communication. A recognition of the fundamental character of interactivity requires a reconsideration of the traditional elements of communication. We are not arguing that the traditional understanding has become irrelevant. Rather, we think that careful and strategic thinking about communication requires us to re-examine all its elements (Table 8.1).

Communication, interactivity, and sociomateriality

The value of being able to work with more traditional communication practices as well as interactivity is illustrated clearly in the case of a large pharmaceutical company (Castelló et al., 2015). This company, which employed a rather traditional organizational communication strategy, experienced new challenges with interactivity after launching a stakeholder engagement platform. As we remarked earlier, interactivity is a general trend in communication, and it had long been important for the organization to maintain its reputation through engagement and dialogue with stakeholders. Now the company wanted to use digital technologies together with workshops and other face-to-face communicative events in order to extend its dialogue with additional stakeholders. They chose Twitter as a platform, but they soon had to acknowledge that the affordances of Twitter did not match traditional communication strategies and practices in the organization. Here, "[the] use of scientific language and terminology ... contrasted with the informal language used in social media platforms; the institutional orientation to hierarchical processes requiring approval for all forms of external communication; and the establishment of fixed working hours that ended at 4pm local time [...] prevented [...] managers from conducting real-time conversations over the Twitter platform" (Castelló et al., 2015, p. 14).

The affordances of Twitter simply did not match the organization's dominant communication practices. Twitter creates expectations about a certain speed and a certain informality in language use. Initially, this led a few employees to break the institutional rules, simply skipping the lengthy and bureaucratic approval process. Later on, they negotiated the rules, initiating the development of a new company communication strategy that allowed for less-controlled conversations between individual employees and external stakeholders. As the senior management expressed it, they needed to recognize that they could control the agenda only part of the time. In these instances, they could put more traditional strategic communication into work. At other times, control over the agenda would be more in the hands of those interacting in the front line. The dual focus on communication and interactivity presented in this chapter, combined with a perspective informed by the theory of sociomateriality, allows us to highlight how more formal communication activities exist alongside new types of interactions, in new discursive and material settings. By new discursive and material settings, we mean that we should take into account more than the channels through which communication travels or the directions of communication flows. We also need to investigate *how* things are expressed; what are the wordings, the images, the plots? And which material elements help shape the communication?

? QUESTIONS

1. How can a given organization work with controlling its image or message while taking into account stakeholders' expectations of interactivity?
2. How do the content and practices on digital platforms impact communication strategies?
3. In what ways do employee communication on digital platforms challenge the traditional organizational divide between the internal and the external?
4. How do live streaming and tweeting impact corporate communication? What are the advantages and pitfalls for organizations that employ these kinds of communication platforms?

References

Albu, O. B. & Etter, M. (2016). Hypertextuality and social media: A study of the constitutive and paradoxical implications of organizational Twitter use. *Management Communication Quarterly, 30*(1), 5–31.

Ashcraft, K. L., Kuhn, T. R. & Cooren, F. (2009). Constitutional amendments: 'Materializing' organizational communication. *The Academy of Management Annals, 3*(1), 1–64.

Asimos, T. (2017). Building a B2B communications strategy around owned, earned, paid and shared. *Circle Studio*, February 22. Retrieved from https://www.circlesstudio.com/blog/building-b2b-communications-strategy-around-owned-earned-paid-shared/

Barnard, C. I. (1938). *The functions of the executive*. Cambridge, MA: Harvard University Press.

Barry, M. & Doherty, G. (2017). What we talk about when we talk about interactivity: Empowerment in public discourse. *New Media & Society, 19*(7), 1052–1971.

Belmondo, C. & Sargis-Roussel, C. (2015). Negotiating language, meaning and intention: Strategy infrastructure as the outcome of using a strategy tool through transforming strategy objects. *British Journal of Management, 26*(1), 90–104.

Berger, P. L. & Luckman, T. (1966). *The social construction of reality – A treatise in the sociology of knowledge*. London: Penguin Books.

Brogaard-Kay, J. (2015). *Constituting performance management: A field study of a pharmaceutical company*. Frederiksberg: Copenhagen Business School.

Bruns, A. (2008). *Blogs, Wikipedia, second life, and beyond: From production to produsage*. New York: Peter Lang Publishing.

Bucy, E. P. (2004). Interactivity in society: Locating an elusive concept. *The Information Society, 20*(5), 373–383.

Carlile, P. R., Nicolini, D., Langley, A. & Tsoukas, H. (2013). *How matter matters: Objects, artifacts, and materiality in organization studies*. Oxford: Oxford University Press.

Castelló, I., Etter, M. & Nielsen, F. Å. (2015). Strategies of legitimacy through social media: The networked strategy. *Journal of Management Studies, 53*(3), 402–432.

Cederström, C. & Spicer, A. (2014). Discourse of the real kind: A post-foundational approach to organizational discourse analysis. *Organization, 21*(2), 178–205.

Chandler, A. D. (1962). *Strategy and structure: Chapters in the history of the industrial enterprise*. Cambridge, MA: MIT Press.

Cheney, G. & Christensen, L. T. (2001). Organizational identity: Linkages between internal and external communication. In F. Jablin & L. Putnam (Eds.), *The new handbook of organizational communication: Advances in theory, research, and methods* (pp. 231–269). Thousand Oaks, CA: SAGE.

Christensen, L. T., Morsing, M. & Cheney, G. (2008). *Corporate communications: Convention, complexity, and critique*. London: SAGE.

Christensen, L. T., Morsing, M. & Thyssen, O. (2013). CSR as aspirational talk. *Organization, 20*(3), 372–393.

Clark, T., Cooren, F., Cornelissen, J. P. & Kuhn, T. (2011). Communication, organizing and organization: An overview and introduction to the special issue. *Organization Studies, 32*(9), 1149–1170.

Conrad, C. & Poole, M. S. (2005). *Strategic organizational communication: In a global economy*. Belmont, CA: Wadworth.

Cornelissen, J. (2014). *Corporate communication: A guide to theory and practice*. London: SAGE.

Costas, J. (2012). We are all friends here: Reinforcing paradoxes of normative control in a culture of friendship. *Journal of Management Inquiry*, 21(4), 377–395.

Czarniawska-Joerges, B. (1997). *Narrating the organization: Dramas of institutional identity*. Chicago, IL: University of Chicago Press.

DeWine, S. (2001). *The consultant's craft: Improving organizational communication*. Boston, MA: Bedford/St. Martin's.

Ewenstein, B. & Whyte, J. (2009). Knowledge practices in design: The role of visual representations as 'epistemic objects'. *Organization Studies*, 30(1), 7–30.

Fairclough, N. (1993). *Discourse and social change*. London: Polity Press.

Fairhurst, G. T. & Putnam, L. L. (2004). Organizations as discursive constructions. *Communication Theory*, 14(1), 5–26.

Frandsen, F. & Johansen, W. (2014). The role of communication executives in strategy and strategizing. In D. Holtzhausen & A. Zerfass (Eds.), *The Routledge handbook of strategic communication* (pp. 229–243). London: Routledge.

Gant, D. & Kiesler, S. (2001). Blurring the boundaries: Cell phones, mobility, and the line between work and personal life. In B. Brown, N. Green & R. Harper (Eds.), *Wireless world: Social and interactional aspects of the mobile age* (pp. 121–131). London: Springer Verlag.

Gheradi, S. & Perrotta, M. (2013). Doing by inventing the way of doing: Formativeness as the linkage of meaning and matter. In P. R. Carlile, D. Nicolini, A. Langley & H. Tsoukas (Eds.), *How matter matters: Objects, artifacts, and materiality in organization studies* (pp. 227–259). Oxford: Oxford University Press.

Gulbrandsen, I. T. & Just, S. (2013). Collaboratively constructed contradictory accounts: Online organizational narratives. *Media, Culture & Society*, 35(5), 565–585.

Gulbrandsen, I. T. & Just, S. (2016). *Strategizing communication*. Frederiksberg: Samfundslitteratur.

Gulbrandsen, I. T. & Just, S. N. (2011). The collaborative paradigm: Towards an invitational and participatory concept of online communication. *Media, Culture & Society*, 33(7), 1095–1108.

Hall, S. (1980). Encoding/decoding. In S. Hall, D. Hobson, A. Lowe & P. Willis (Eds.), *Culture, media, language* (pp. 128–138). London: Routledge. (Original work published 1963.)

Hallahan, K., Holtzhausen, D., van Ruler, B., Verčič, D. & Sriramesh, K. (2007). Defining strategic communication. *International Journal of Strategic Communication*, 1(1), 3–35.

IKEA (2019). IKEA history - how it all began. Retrieved at https://www.ikea.com/ms/en_AU/about_ikea/the_ikea_way/history/

Jenkins, H. (2006). *Fans, bloggers and gamers: Exploring participatory culture*. New York: New York University Press.

Kiousis, S. (2002). Interactivity: A concept explication. *New Media & Society*, 4(3), 355–383.

Klikauer, T. (2007). *Communication and management at work*. London: Palgrave Macmillan.

Koushik, S., Birkinshaw, J. & Crainer, S. (2009). Using Web 2.0 to create Management 2.0. *Business Strategy Review*, 20(2), 20–23.

Lasswell, H. (1948). *The structure and function of communication in society. The communication of ideas.* New York: Institute for Religious and Social Studies.

Leonardi, P. M., Huysman, M. & Steinfield, C. (2013). Enterprise social media: Definition, history, and prospects for the study of social technologies in organizations. *Journal of Computer-Mediated Communication, 19*(1), 1–19.

LoBianco, T. & Raju, M. (2017). Intel Dem: Trump may have declassified CIA hack information. *CNN Politics*, March 16. Retrieved from http://edition.cnn.com/2017/03/16/politics/adam-schiff-donald-trump-hack

Manuti, A. (2016). Communicating the 'social' organization. In A. Manuti & P. D. de Palma (Eds.), *The social organization: Managing human capital through social media* (pp. 14–27). Basingstoke: Palgrave Macmillan.

Marrewijk, A. V. & Yanow, D. (2010). Introduction: The spatial turn in organizational studies. In A. van Marrewijk & D. Yanow (Eds.), *Organizational spaces* (pp. 1–16). London: Edward Elgar Publishing.

May, S. & Mumby, D. (2005). *Engaging organizational communication. Theory and research: Multiple perspective.* Thousand Oaks, CA: SAGE.

Mazmanian, M., Orlikowski, W. J. & Yates, J. (2013). The autonomy paradox: The implications of mobile email devices for knowledge professionals. *Organization Science, 24*(5), 1337–1357.

Morley, D., Shockley-Zalabak, P. & Cesaria, R. (2002). Organizational influence processes: Perceptions of values, communication and effectiveness. *Studies in Communication Sciences, 2*(1), 69–104.

Orlikowski, W. J. (2007). Sociomaterial practices: Exploring technology at work. *Organization Studies, 28*(09), 1435–1448.

Phillips, L. (2011). *The promise of dialogue: The dialogic turn in the production and communication of knowledge.* Amsterdam: John Benjamins Publishing.

Plesner, U. & Gulbrandsen, I. T. (2015). Strategy and new media: A research agenda. *Strategic Organization, 13*(2), 153–162.

Putnam, L. L. & Nicotera, A. M. (2009). *Building theories of organizations: The constitutive role of communication.* New York: Routledge.

Riel, C. B. M. V. & Fombrun, C. J. (2007). *Essentials of corporate communication: Implementing practices for effective reputation management.* Oxon: Routledge.

Schein, E. H. (1985). *Organizational culture and leadership.* San Francisco, CA: Jossey-Bass.

Sifry, M. L. (2011). *Wikileaks and the age of transparency.* New York: OR Books.

USAjobs.gov (2016) Information technology specialist – Department of Justice. Retrieved from: https://www.usajobs.gov/GetJob/ViewDetails/452934900/

Villadsen, K. (2017). Constantly online and the fantasy of 'work–life balance': Reinterpreting work-connectivity as cynical practice and fetishism. *Culture and Organization, 23*(5), 363–378.

Weick, K. E. (1995). *Sensemaking in organizations.* Thousand Oaks, CA: SAGE.

9 LEGITIMACY AND TRANSPARENCY

> This chapter describes how legitimacy has become a central concern for organizations, and how demands for more transparency are linked to legitimacy. It argues that expectations of transparency have grown with the advent of digital communication technologies. The chapter opens with a historical account describing how it has become increasingly important whether an organization and its activities are viewed as legitimate from a societal point of view. It then introduces two classical approaches to legitimacy, namely one that holds that the legitimacy of an organization is a matter of collective negotiation, and a more practically oriented approach, by which organizations can manage legitimacy. After this basic introduction to how legitimacy has been approached in organization studies, the chapter moves on to describe how legitimacy is increasingly tied to transparency. We zoom in on the concept of transparency first by discussing the simple understanding of transparency as a practice of delivering 'more and better information'. We then move on to a discussion of how transparency is 'manufactured' in practice. By looking at different methods of producing transparency, we show how humans and material or technological entities play a role in manufacturing transparency in particular ways. From our discussions throughout this chapter, it becomes clear that legitimacy is not something organizations 'have' and that transparent is not something organizations 'are'. Both these phenomena are dependent on both people and technologies. The central argument of this chapter, then, is that digital technologies affect how organizations construct their legitimacy and transparency. We treat these two as interconnected phenomena because digital technologies create new expectations to – and means of producing – transparency.

Why legitimacy and transparency?

With the increasing significance of organizations in society (Perrow, 1991), the question of whether an organization and its activities are legitimate from a societal point of view has come to the forefront. One aspect of legitimacy is transparency. If we are to judge whether an organization is behaving according to societal expectations, it seems evident that we need to be able to look over the shoulders of its managers and employees. With the advent of digital technologies, it has become easier to share information and to gain access to information about organizations, whether they like it or not. Against this background, we are witnessing a general rise in expectations regarding how organizations interact legitimately with the surrounding society, and how they disclose information that allows the public to judge whether they act legitimately. Obviously, there are organizations such as intelligence agencies, organized crime networks, or controversial associations where it is expected or accepted that they are non-transparent (Scott, 2013), but here we will focus on the general trend.

Legitimacy and transparency: The case of an oil company

Let us consider the example of an organization that made use of digital communication platforms to work strategically with legitimacy and transparency. The organization – the Brazilian state-owned oil company Petrobras – had a long history of being challenged by various publics, and had been subjected to negative media coverage of its activities (for the full story, see Barros, 2014). The mass media have traditionally played an important role in relation to organizational legitimacy as they "are able to blame, applaud, and scandalize, and in doing so, have the potential to (de)legitimize managerial and organizational behavior" (Hartz & Steger, 2010). Petrobras decided to establish a corporate blog in order to regain the legitimacy that was continuously being challenged by mass media outlets. A blog obviously allows an organization to offer its own perspective on its activities and its place in the world, and it does so in a dynamic fashion, offering the opportunity to continuously update and comment on events as they happen and on issues as they emerge.

The blog strategy followed two main routes: one was to challenge media text representations, and the other was to demonstrate transparency through sharing large amounts of information. Both routes led to a successful destabilization of the legitimacy of traditional media's control of information, as well as the media's alleged partisanship. Specifically, the organization used the blog to systematically deconstruct media texts, responding to reporters' accusations. In doing so, they applied a formal and neutral rhetoric, and, importantly, they worked strategically on appearing transparent. Petrobras stated that they wanted to make all their answers to the press available "without data editing" (Barros, 2014), so the public could judge who was most legitimate – the company or the press. They also inserted hyperlinks to original sources of information so that readers could pursue an issue and see that other sources were backing the company's claims. Finally, Petrobras sought to expose what they saw as traditional media's 'ideological bias' by publishing very diverse newspaper headlines revolving around the same event. The organization allowed the headlines to speak for themselves, and this managed to produce a lot of engaged comments from the public such as "Oh, my god! You only have to read the headlines to see the editorial opinions of each newspaper. It's astonishing how they manipulate the information!!!" (Barros, 2014, p. 1222).

The Petrobras media case shows that legitimacy, and the strategies to establish, maintain, or repair legitimacy, can be a serious matter of concern for an organization. It also shows how legitimacy and transparency are tied together, and that digital technologies play a role in working on both. In the following section, we will highlight some important insights from organization studies regarding legitimacy and legitimation strategies. We will then argue that with digital technologies, the possibilities to work strategically with legitimacy are increasingly tied to transparency. We do not argue that transparency is simply replacing other types of legitimation work. However, where digitalization leads to challenging new questions about legitimacy, the response seems to include a move toward transparency.

Legitimacy as a theme in organization studies

When organizational legitimacy began to emerge as a topic of scholarly discussion, it was viewed largely in economic and legal terms. Milton Friedman (1962, p. 133) famously stated that: "there is one and only one social responsibility of business – to use its resources and engage in activities designed to increase its profits so long as it stays within the rules of the game." This way of thinking has been called 'the classical model of corporate legitimacy' (Grolin, 1998). It was closely related to early capitalism and demonstrated that the primary concern for corporations was the satisfaction of stockholders' desire for profitability. As Friedman (1970) said, corporations should accept no other responsibilities than that of making as much money as possible for their shareholders.

Although these ideas are still articulated – and still guide the actions of many companies – they have been challenged by societal changes. It has become less acceptable to articulate shareholder profit as the only thing that matters to a corporation. From the 1930s, both labor conflicts and the gradual realization that societies were deeply affected by corporate choices and actions led to new demands on corporations. This can be captured by what has been referred to as 'the stakeholder model of corporate legitimacy' (Grolin, 1998). This model "emphasizes the obligation of corporations to be responsive to those in society immediately affected in a tangible way by corporate decisions and actions" (Grolin, 1998, p. 216). Those affected can be employees, local communities, business partners, and other social groups. Taken together, they are referred to as *stakeholders*, a development of – or play on – the word *stockholders*. The terminology indicates the need to take into account both those who have stocks in a corporation and those who experience having something 'at stake' in relation to the corporation. Below, we will discuss stakeholder theory and strategies for stakeholder engagement in more detail, as this model is still widely used as a framework for working with organizational legitimacy. But as a response to increasing demands on organizations, a more comprehensive model has also emerged: 'the political corporation model'. As Jesper Grolin (1998, p. 217) describes it:

> The political corporation model, reflecting an increasing globalization of the economy and a parallel weakening of national governmental authority, goes several steps further by demanding that a corporation adopts a clear set of moral and ethical values, which relate to the general public, both locally and globally, and which guide corporate actions irrespective of whether it is explicitly required by law or demanded by the public.

This diagnosis goes some way toward explaining the amount of effort and resources organizations currently spend on establishing or maintaining legitimacy. The expansion of this model has become possible because there is wide agreement in society that the responsibility of organizations goes beyond pursuing their immediate interests (see Table 9.1 for an overview of the three models). As important societal entities, organizations cannot ignore the sentiments expressed in the third model. In the organization literature, legitimacy has been described as a valuable asset because it enhances the stability and comprehension of organizational activities. Conversely, organizations that lack legitimacy are vulnerable to criticism and to "claims that they are negligent, irrational or unnecessary" (Meyer & Rowan, 1977, p. 350). In the following sections, we will look more into the literature that explains this relationship between societal expectations and organizational legitimacy work.

Institutional approaches: Legitimacy as evaluative process

Today, organizations do not have an unquestioned 'license to operate' in society. They are dependent on societal demands and expectations. Not a day passes without organizations being criticized for violating not only legal

Table 9.1 Models of corporate legitimacy (based on Grolin, 1998).

The classical model	The stakeholder model	The political corporation model
Responsibility toward shareholders and the law	Responsibility toward those affected by corporate actions	Responsibility toward the society in which the corporation operates

demands and financial regulations, but also ethical standards. Banks are held accountable for helping their customers evade taxes, tobacco companies and fast-food chains for selling unhealthy products, and tech giants like Facebook for treating their users' data unethically or, in the case of Google, for unfair competition. Sometimes legal actions are taken, but in other instances, organizations react to public opinion, for instance by apologizing for being insensitive to a minority group or by promising to 'listen' to users, consumers, or the community. Their perceived need for legitimacy is a driver of organizational change when they choose to adapt to pressure from their environment. Why, then, have corporations become so sensitive to these outside pressures? What explains these dynamics? According to Mark Suchman, "sector-wide structuration dynamics generate cultural pressures that transcend any single organization's purposive control" (Suchman, 1995, p. 572). Put differently, there are simply forces outside the organization which are much stronger than managerial decisions. Suchman captures the nature of these forces in his definition of legitimacy. From his perspective, legitimacy is "a generalized perception or assumption that the actions of an entity are desirable, proper or appropriate within some socially constructed system of norms, values, beliefs, and definitions" (Suchman, 1995, p. 574).

By 'generalized', Suchman points out that the legitimacy of an organization is a matter of collective negotiation, not single viewpoints. Referring to socially constructed systems of norms, values, beliefs, and definitions, he adds a discursive perspective to the question of legitimacy. As we discussed in Chapter 8 on communication and interactivity, some of the organizational literature views organizational phenomena as being discursively constructed. This is also the case with the 'institutional' approaches to legitimacy, of which Suchman's work is an example. From this perspective, legitimacy is not something an organization 'has', but something continuously negotiated and eventually lent to the organization by societal actors. If we take the example of banks again, the dominant discourse surrounding large, traditional banks used to take their mode of operating for granted or even portray them in a respectful light; after the financial crisis, banks are referred to in much more critical terms – for instance, coupled with concepts such as 'unhealthy culture', 'sick incentives', or the like.

In organizational discourse studies, it has been shown how discursive struggles and the consolidation of particular discourses are central to achieving a legitimate position. The point of discourse analysis is precisely that discourses are not isolated instances of language use. Rather, discourse must be studied in a wider setting than within an individual organization. For instance, when Eero Vaara and colleagues (2006) studied the different discursive strategies used to make sense of – and legitimize – a controversial merger, they went beyond the individual organizations and saw how legitimation processes took place in the media – or in a collective setting, as it were. They identified a range of discursive strategies that when pursued gradually made the merger seem legitimate.

Although managers can use discursive resources strategically, and probably do so with some effect, this approach does not allow for much managerial agency in relation to organizational legitimacy. As mentioned, the legitimacy of an organization is a matter of collective negotiation. In that way, the task of managing stakeholders becomes daunting. The view on managerial agency is different in another stream of literature, namely the strategic approaches to legitimacy.

Strategic approaches: Legitimacy as a goal

In the strategic approaches, legitimacy is conceived of as an operational resource. Strategic approaches are managerial in the sense that they assume a high level of management control over the legitimation process. Calling legitimacy 'strategic' emphasizes how organizations can instrumentally work to garner societal support. This approach to legitimacy has entered the field of strategic management. For instance, the much-used and much-quoted strategy scholar Michael Porter has also theorized how managers can work strategically with legitimacy – by working with Corporate Social Responsibility (CSR) in a very particular way. According to Porter, the dominant logic in CSR should be that corporations and society create shared value (Porter & Kramer, 2011); corporations should not blindly engage in philanthropy in order to appear legitimate. Rather, they should examine their value chain to see where it makes sense for the business to solve societal problems. Companies have rushed into this new way of doing CSR by, for instance, making their transportation less energy-consuming and thereby reducing CO_2 emissions, or by educating producers of raw materials in creating better products, thereby making a difference for local communities and obtaining access to more valuable resources. The idea is that when local community members are trained in, for instance, new production methods, this benefits both them and the company that purchases their products.

This type of CSR strategy raises ethical questions as to why businesses engage in the solution of societal problems – is it only to *appear* legitimate? Is it to *become* legitimate? Or, is it simply because it is good for business? When such approaches are described in the literature, it is often in less than positive terms. For instance, Suchman talks about strategic manipulation as "the ways in which organizations instrumentally manipulate and deploy evocative symbols in order to garner societal support" (Suchman, 1995, p. 572). However, Suchman also recognizes that organizations face real issues in relation to legitimacy and that they need to relate strategically to legitimacy challenges. He has described how legitimacy management can take place in different situations: when organizations need to *gain* legitimacy, when they need to *maintain* legitimacy, and when they need to *repair* legitimacy (Suchman, 1995).

Gaining legitimacy is an issue when an organization embarks on a new activity or is first established. For instance, organizations in new industries like business coaching need to work hard to establish legitimacy both internally and externally (Clegg et al., 2007). In this situation, the new organization faces challenges such as conforming to the environment or creating new constituencies. Maintaining legitimacy is depicted as less challenging, because legitimacy, once conferred, tends to be taken for granted. However, an organization has many different stakeholders, who might have conflicting views of its activity. Moreover, the organizational context might change over time. Therefore, two strategic necessities consist in recognizing future changes and protecting past accomplishments. One example could be fitness centers, which continuously adjust to health-related research by closing sun lounges, banning the use of steroids, introducing more natural snacks, or bringing in the latest exercise regimes. The most forward-looking fitness chains have held on to their original workout ideals, but they have also evolved in tandem with the expectations of their users and the norms of the surrounding society. Finally, the task of repairing legitimacy involves both reacting to a crisis in trust, typically some kind of scandal, and in re-establishing legitimacy based on new grounds – since otherwise established practices are no longer serving their legitimizing purposes. Volkswagen is a good example. They trumpeted the low emissions of their diesel cars, but it was revealed that they used software in the cars that helped them cheat on emissions tests. VW acknowledged that they had "broken the trust of [their] customers and the public" and that their most urgent task was to win this trust back (through internal investigations and the firing of the guilty engineers and executives) (Hotten, 2015, n.p.). Legitimacy and trust are intertwined – a company that loses legitimacy is likely to lose trust as well.

As we see, even if legitimacy is presented as a strategic goal, it is not an issue that can be solved once and for all. The important lesson for organizations seems to be that they need to take legitimacy seriously and understand its multiple dimensions: "Because real-world organizations face both strategic operational challenges and institutional constitutive pressures, it is important to incorporate this duality into a larger picture that highlights both the ways in which legitimacy acts like a manipulable resource and the ways in which it acts like a taken-for-granted belief system" (Suchman, 1995, p. 577).

Organizations must therefore have the capacity to analyze their social context and to address different constituencies communicatively. We will argue that digital technologies make a difference in both these areas. They help shape new conditions for organizations' understanding of their social context, and they afford new ways in which organizations can communicate/interact with different constituencies. With the internet, we can no longer talk about a public sphere where a small number of closely edited media control meaning formation as well as the types of issues raised in public. Instead, individuals and interest groups are able to bring up a large number of problems, stories, and interpretations, put these issues 'out there' and compel organizations to react. For instance, pharmaceutical companies need to take into account a myriad of stories of side effects, complaints about price levels, and criticism from proponents of alternative medicine. With the advent of digital technologies and unavoidable interactivity (see Chapter 8), organizations need to rethink their communication strategies, taking into account the specificities of online communication dynamics (Gulbrandsen & Just, 2013).

Stakeholder theory

One practice-oriented approach to managing legitimacy has developed under the rubric of 'Stakeholder Theory'. Stakeholder Theory emerged from a concern with the increasing difficulty of managing a corporation, given that corporations face growing amounts of government regulations, criticism, competition, and so on. As R. Edward Freeman (1984, p. v) puts it: "Managers in today's corporation are under fire. Throughout the world, their ability to manage the affairs of the corporation is being called into question ... He or she finds an increase in the external demands placed on the corporation and a decrease in the internal flexibility of the corporation".

The concept of 'stakeholder management' is then offered as an alternative way of thinking about the management challenge. The challenge can be broken into smaller parts by defining who stakeholders are, and having a strategic approach to each stakeholder group. By definition, a stakeholder is "any group or individual who can affect, or is affected by, the achievement of a corporation's purpose" (Freeman, 1984, p. vi). It has been pointed out that this is a rather imprecise point of departure for strategic work, and several frameworks have been offered to systematically analyze who the most important stakeholders are; that is, those stakeholders on whom an organization should spend its resources. For instance, Ronald Mitchell and colleagues (1997) proposed that stakeholder management should focus on those stakeholders (1) with most power, (2) who have legitimate claims in relation to the organization, and (3) whose demands are most urgent. Most models include an assessment of how much *influence* a stakeholder has on the organization, and how much *interest* the stakeholder has. Based on an analysis of these factors, stakeholder management can prioritize those who have most influence and highest interest, and otherwise monitor the remaining groups who have less influence and less interest.

Since the context and the stakeholders of an organization are dynamic entities, an enduring, authoritative mapping is hard to make. Hence, effective stakeholder management must be sensitive to changes and complexities in the organization's environment. This task becomes all the more pressing with digitalization. Although digital communication platforms can certainly be used to reach out to stakeholders, they also render stakeholder management much more complex. For instance, it has been argued that open social media platforms such as Twitter are suitable for engaging in dialogue with stakeholders because they seem to be symmetrical and without gatekeepers (Castelló et al., 2016). At the same time, open platforms also increase the complexity of communication because they lay bare the multiple and conflicting agendas of stakeholders. Stakeholders are no longer neatly defined entities. They must now be addressed in relation to their changing affinities. In this context, the dialogic features enabled by webpages, blogs, and social media have proven to be difficult to apply in interaction with stakeholders (Castelló et al., 2016), as we also argued in Chapter 8.

In their study of Twitter communication, Itziar Castelló and colleagues describe how much work employees need to put into understanding stakeholders' needs if they are to respond appropriately. Employees "realized that entering a conversation required gaining acceptance as part of the community". They also discovered that "'there are agendas by a variety of stakeholders of which we are not aware of, and where we might want to join" (Castelló et al., 2016, p. 417). It delegitimizes the organization's communication if employees do not succeed in identifying people's concerns. In this way, digital platforms lead to shifting power dynamics, because stakeholder communication is not contained in offline meetings or other closed settings, but becomes transparent in all its messiness – potentially challenging any orderly image of the organization and its activities. This accentuates the importance of understanding the interactivity afforded by digital technologies, as discussed in the chapter on communication and interactivity.

Transparency

We already saw in the above that an organization's legitimacy work often involves attempts to become or appear more transparent, and that digital technologies are an important element in this work, sometimes in productive ways, sometimes in disturbing ways. However, we should stress here that transparency has been an organizational issue long before the explosion in digital technologies.

More and better information

Ideals about transparency can be traced back to the Enlightenment, and transparency is often described as a desire for clarity, insight, and participation (Christensen & Cheney, 2015). In organizations and management, transparency has become an increasingly inevitable issue to tackle. Consider the establishment, in 1993, of the global civil-society organization Transparency International. It is first and foremost the institutionalization of an anti-corruption movement, but its primary value is transparency (Transparency.org, 2019). The vision of Transparency International is to *promote transparency, accountability and integrity at all levels and across all sectors of society.* So, in principle, no organization is spared scrutiny – and we have also witnessed how whistleblowers have managed to appeal to transparency as an ideal while creating serious problems for various organizations (see du Plessis, 2014; Perry, 1998). If we consider the major scandals involving the leaking of vast amounts of classified material via digital platforms (think of Edward Snowden's leak from US intelligence archives or the Panama Papers scandal, where millions of confidential documents from a law firm helping tax evaders were made public), these can be considered elements of the 'transparency movement' (Sifry, 2011).

Transparency is also considered a virtue when it comes to the internal operations of organizations. Ostensibly, transparency is necessary in order to avoid a number of organizational dysfunctions. As one management consultant expresses it: "Typical symptoms of lack of transparency are sub-optimization, duplicate work, bad decision-making and inability to innovate. These things don't just impact the bottom line results negatively – they also hamper an organization's ability to compete and survive in the long haul" (Berg, 2011, n.p.).

In organization studies, there is a growing interest in transparency as a theme (Bernstein, 2017), associated with the growing interest in disclosure, visibility, monitoring, surveillance, and so on. From a scholarly perspective, transparency is not a single, demarcated process or concept. There are many nuances in how we can understand organizational transparency. For instance, as Ethan Bernstein suggests, transparency might be physical (as in shared office spaces) or with regards to data sharing; it can be immediate or delayed (making decisions where disclosures are made after the fact); it can be consensual or mandated (compensating employees for sharing versus forcing them to do so); or it can be anonymous (as in apps such as Whisper, where information can be shared anonymously) or identifiable (Bernstein, 2017). In organizational life, recommendations are less nuanced and very ambitious. For instance, if we check the online business dictionary, we realize that for organizations, transparency is a difficult ideal to fulfill, as it implies 'lack of hidden agendas' – 'availability of full information required for collaboration' – 'minimum degree of disclosure', and that 'rules and that reasons behind regulatory measures are fair and clear to all participants' (Businessdictionary.com, 2019).

Linking back to our discussion of stakeholders above, the literature on Public Relations has offered guidelines both for stakeholder communication (which should be based on 'complete and reliable information') and for evaluating how stakeholders judge the transparency of an organization (Rawlins, 2008). In the corporate communication field as well, transparency is considered an important precondition for legitimacy (Christensen & Cheney, 2015). More and better information is assumed to make the organization appear more trustworthy and responsible, thereby making them legitimate actors in society. As expressed in the corporate reputation literature, "transparency is at the heart of public trust and represents the underbelly of corporate reputation" (Fombrun & van Riel, 2004, p. 94).

More and better information in practice: The case of digital learning platforms

Let us turn to an example of how 'more and better information' is put to use in a public school context as a legitimation strategy based on the idea of transparency. It can be argued that any public school needs to continually contribute to the reputation of the public school system and take initiatives to uphold a legitimacy otherwise threatened by critical taxpayers, politicians, parents, and media scandals over testing, inclusion, or disorderly pupils. Digitalization strategies and projects can thus be seen as legitimation work, which is pursued by different actors in the public school system both to demonstrate that public schools are dynamic and oriented toward quality, and to become or appear more transparent. At the most basic level, many public schools have embraced digital technologies as replacements for analogue media; smart boards are introduced to the classroom, students use their phones for research on school projects, and books are sometimes replaced with digital material. Besides the digitalization of the classroom, other aspects of organizing the school are digitalized via various digital learning platforms. In the case of Denmark, as in many other countries, the national and local governments impose the use of learning platforms on all public schools to "support learning as well as a flexible planning and execution of teaching independent of time and place" (KL, 2015, n.p.).

A central feature of the learning platform is that students, parents, teachers, and school administrators have access to developmental plans and portfolios for the individual pupil, as well as teaching plans, learning objectives, and teaching materials. All these features create a high degree of transparency in teaching and learning – administrators, classroom teachers, parents, and pupils can obtain a detailed picture of what takes place in the classroom. The classroom had otherwise traditionally been the domain of an independent classroom teacher, so the advent of the learning platform and digitalization can be seen as a radical change in teachers' working conditions. It also demands that a lot of documentary material be produced for the purpose of sharing. Digital platforms thus afford the creation of a new kind of transparency – and produce new organizational phenomena such as increased possibilities of control, collaboration, communication – and management. As we saw in Chapter 1, in the example of the school teacher who 'got fired by an algorithm', registrations and data are used extensively to render teaching more transparent as part of 'quality work'. The flipside to this system is when technologies are *non-transparent* (recall our discussion of spaghetti code in Chapter 1) and lead to illegitimate decisions.

A similar phenomenon can be seen in the case of surveillance technologies, especially when they are no longer solely used for crime prevention or security, but are now brought into organizations in order to monitor employees at a distance. For instance, some day care centers have installed

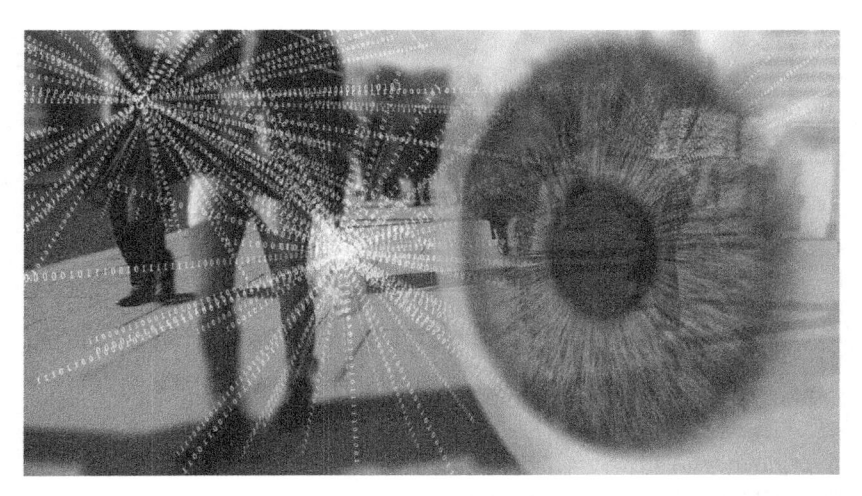

Image 9.1 Our digital devices leave traces even when we move through the city. This creates new surveillance issues: When are we being watched, and by whom?

Source: Peterhowell, iStock.

surveillance cameras so as to allow parents to access footage in real time on the internet. This establishes new working conditions for caregivers, who do not know when they are being observed, and it potentially establishes new relationships between parents and caregivers if the latter need to defend a range of small events observed by parents. Apparently, some organizations see these developments as a possibility to display their strengths ('we have nothing to hide'), while others problematize it as an Orwell-esque 'Big Brother' phenomenon (Lombardi, 1997). The discussions about transparency versus privacy have been rather isolated from one another in the academic literature (Bernstein, 2017). But, it has been argued that people willingly interact with technologies for their own purposes even if they are tracked and data about them is collected (Image 9.1). In this view, surveillance technologies are not to be seen as solely exploitative nor solely benign (Ellerbrook, 2010).

The cases, the managerial advice, and the literature we have discussed here all revolve around the ideal of 'more and better information'. In organization studies, however, an interest has emerged in how transparency is 'manufactured'. As Hans K. Hansen and Mikkel Flyverbom (2015) point out, scholarly discussions about transparency originally focused on the amount and quality of information shared with the public, but have evolved into an interest in the social dynamics of negotiating accountability (which we will consider part of legitimacy here). This trend overlaps with the growing recognition that the one-way transmission of information does not really satisfy the demand for transparency. This is because any simple model of transmitting information leaves us blind to the power involved in selecting and presenting information, and it cannot account for how specific technologies mediate transparency in specific ways (Hansen & Flyverbom, 2015). In the coming section, we will focus on these issues.

Transparency and mediating technologies

As we have seen, it is common to talk about transparency as a *principle* that generates trust (or at times, a principle that generates suspicion), rather than a *practice* that depends on technologies. But while decisions about what to disclose and conceal have always been an integral part of organizational life, debates about transparency have been invigorated with the advent of ubiquitous digital technologies (Flyverbom et al., 2016). In this section, we will discuss both the organizational manufacturing of transparency and transparency as unwanted exposure. We will then relate these aspects of transparency to digital technologies.

Organizational manufacturing of transparency

If we look at transparency as an organizational *practice*, it is possible to identify a range of different methods that organizations employ to *produce* transparency (Hansen & Flyverbom, 2015). Hansen and Flyverbom identified three very different types of methods, which they call 'qualitative due diligence', 'quantitative rankings', and 'algorithmic big data analysis'. Qualitative due diligence is the practice of examining organizational phenomena through interviews, meetings, peer reviews, background checks, reading business histories, consulting databases, and similar methods requiring people's involvement and judgment. Hansen and Flyverbom describe how such methods are frequently used in preparing for mergers and acquisitions, where the right information – and right amount of information – is very important for the decision-making process. In these qualitative due diligence methods, it is relatively easy to see that transparency is the product of human efforts to gain 'a full picture' of an organization. This is less so in the second set of methods, which they call 'quantitative rankings'.

It is now common to use rankings and benchmarks as indicators of organizational performance. To create rankings or to use benchmarking, it is necessary to establish quantitative baseline measurements which allow for comparisons between phenomena or organizations that might be otherwise heterogeneous in practice. When messy organizational phenomena are grouped, quantified, and measured in relation to one another, an aura of transparency is created. We say 'aura' because these quantification processes often conceal why and how particular elements came to count in the measurements. Wendy Espeland and Michael Sauder (2007) describe the extensive organizational consequences of this type of transparency production: When doctors' performances are publicly ranked, their professional decisions begin to be influenced by how their performance scores will look, and when law schools are publicly ranked, this ranking influences the number and quality of applications they receive and the financial resources they are able to obtain in the future. In this sense, rankings do not just objectively depict the quality of medical treatment or legal education, they also shape these phenomena. Proponents of rankings insist that they offer valuable objective data for users, enabling users to make judgments. The more critical voices point to the counterproductive practices arising from the quantification mania and the compulsion to produce this type of transparency.

The final type of transparency production is 'algorithmic big data analysis'. In Chapter 7 on knowledge and datafication, we discussed the increasing significance of big data analysis in organizations. Big data is also an issue in relation to transparency, because it appears on first sight that the automated generation of massive data sets can produce knowledge about actual practices, and that this knowledge can be teased out by computers. However, Hansen and Flyverbom (2015, p. 883) sound a note of caution:

> The notion that 'numbers speak for themselves' echoes our earlier discussion of transparency understood as unmediated access to reality. Here it is complemented with the belief that large data sets can generate insights that were previously impossible and 'with an aura of truth, objectivity, and accuracy' (boyd & Crawford, 2012, p. 663). But the access to reality is mediated by algorithmically coded soft and hardware devices, which afford particular kinds of knowledge and insights, but never the full picture of anything.

The point is that we fool ourselves if we believe that big data offers a neutral depiction or a 'full picture' simply because it measures phenomena quantitively on a large scale. In relation to transparency, the problem with knowledge production based on automated big data analyses is that such analyses are detached from real-life situations and practical experience. It becomes difficult to judge the validity of what is being disclosed, and the seemingly objective transparency becomes difficult to challenge because the mode of producing it is so technical. There are several examples of big data analyses produced by Google or Facebook that are so opaque as to make them useless. For instance, Google Flu Trends can potentially be used by doctors and health agencies to spot disease outbreaks, but users and scientists have no way of knowing how the flu data is produced, as Google will not reveal its algorithm. David Lazer and colleagues (2014) showed that although Google has massive amounts of data, the predictions they produce are not necessarily accurate. Unlike normal scientists, Google is so secretive that it is impossible for independent researchers to even try to replicate their results or improve their methods for handling the data (See Table 9.2 for an overview of methods for manufacturing transparency.).

DIGITAL ORGANIZING

Table 9.2 Methods for manufacturing transparency (based on Flyverbom & Hansen, 2015).

Qualitative methods	Quantitative methods	Big data
Transparency generated by interviews, meetings, documentary studies	Transparency generated by rankings and benchmarks	Transparency generated by automated processing of large data sets

The point of looking into different methods of producing transparency is that it exposes how humans and technological entities play a role in manufacturing transparency in particular ways. This approach not only problematizes the ideal of transparency. It offers an understanding of the organizational work that needs to be done to work with transparency.

Transparency as unwanted exposure: The case of TripAdvisor

From the point of view of the literature we have discussed here, transparency is not just a straightforward matter of disclosing more and better information. Transparency is manufactured by organizations and mediated in specific ways by technologies. However, besides this organizational work with transparency, there are also dimensions of transparency, which arise in the interplay between technologies and a multitude of actors who are *external* to the organization. These external relations have massive implications for the organization and its operations. The case of TripAdvisor illustrates nicely how digital technologies sometimes have agencies that escape managerial control, and that transparency can transcend even the most carefully conceived organizational strategies. TripAdvisor, the website for hosting online user reviews of hotels, is based on the idea that the best travel advice comes not from guidebooks or corporate websites. Instead, it comes from uncensored evaluations by 'the crowd', or as the company frames it, "unbiased content from real travelers" (Orlikowski & Scott, 2014, p. 881). Wanda Orlikowski and Susan Scott studied the different practices involved in evaluating hotels in both the traditional hotel star classification practice and on this crowd-based internet site. They had a specific focus on the interactions between businesses and evaluators, and on the organizational implications of the different types of evaluation. The case of TripAdvisor can be interpreted as an example of the transparency movement aimed at consumers. It is also framed as an initiative to create more transparency on the basis of which consumers can act. For instance, an employee at TripAdvisor says: "Transparency is the word ... So with us, we put all the reviews online and they are not edited by us, they are totally what the person is saying. It's really important for us to be trustworthy and to make sure nothing is changed and everything is transparent" (Orlikowski & Scott, 2014, p. 881).

TripAdvisor has more than 400 million unique visitors per month, so the popularity of this type of online transparency is evident. It has had a dramatic impact on the hotel industry. For some of the hotels displayed on the site, their new online visibility has entailed a radical increase in bookings and income. These hotels have become less dependent on 'expert judgment' and guidebook selections. For other hotels, however, being exposed to travelers' opinions and complaints has resulted in an increased vulnerability and loss of control. Subjective opinions of travelers and revealing photos (stained carpet, dirty bathroom) are not manageable, and they remain accessible online for years to come. Furthermore, although hotel owners can and do respond to guests' comments, it is not possible to 'speak back' to the TripAdvisor organization, as much of the ranking work is done with the help of an algorithm (and it is not transparent how that works). And since TripAdvisor prides itself on not interfering with content, it is an open question how hotels should deal with fake or unjust reviews. Even if it is obscure how evaluations and rankings are produced, and even if their validity can be challenged, hotels have no choice but to act on them. As Orlikowski and Scott (2014, p. 883) write:

> Hoteliers find themselves obsessing over the flow of current opinion about their operations as it streams into their email inboxes every day, nudging and activating them to respond. This prompts them to contingently and continually improvise their practices in an attempt to maintain or improve their current rank on the Popularity Index, while managing the multiple, divergent, and often-contradictory evaluations they are confronted with on a regular basis. Their practices are bound to the ranking algorithm and 'micro-managed by the crowd'.

When reviews are anonymous and algorithmic operations are non-transparent, when more information does not necessarily mean better information, there are no simple ways of ensuring that the reviews do not damage one's business. Therefore, organizational initiatives revolve around attempts to integrate user-generated input into management. For instance, Orlikowski and Scott saw that TripAdvisor comments were used in staff meetings and in decisions to fire employees, and that this type of intervention in management by a crowdsourcing platform "produces hotel staff who feel that they are under constant surveillance; every guest is their boss, there is no opportunity for recovery, they are ever anxious" (Orlikowski & Scott, 2014, p. 886). In the case of TripAdvisor, we thus see two aspects of transparency in play. We see the celebration of transparency; the idea that users can get 'the real picture' based on a type of crowd wisdom. But we also

Image 9.2 Crowdsourcing platforms challenge organizations by exposing them to uncensored criticism. Organizations may try to speak back to them in various ways.

see a lack of transparency; that it is impossible to question how knowledge is generated or judge the validity of that knowledge. Orlikowski & Scott argue that whereas traditional hotel reviews were based on clearly articulated criteria, the opposite is the case for judgments made by individuals composing a crowd. In any case, agency is moved from the organization to digital technology and the crowd. Not in the sense that the hotel has no agency, but in the sense that the crowd defines what the hotel needs to react to.

Legitimacy, transparency, and sociomateriality

From our discussions throughout this chapter, it has become clear that legitimacy is not something organizations 'have' and transparent is not something organizations 'are'. Both phenomena are dependent on people and on the technologies they employ. The examples allowed us to see that a given technology holds promises and also poses potential problems in relation to legitimacy and transparency. The role of any given technology cannot be determined in advance. For organizations, this implies that attention must be given to the interconnectedness of accountability practices and digital technologies. An interesting case is the German Pirate Party, which has become famous for articulating an ideology of transparency. The party considers it "a right for every member to express their opinions publicly without any restriction and a moral obligation for party officials to document decision-making processes as well as all discussions in councils or committees" (Ringel, 2019, p. 7). Hence, the Pirate Party video-streams or audio-tapes official meetings and discussions, they make interactions between meetings public, each document produced by working groups must be published online, and mailing lists are open to every citizen. The story goes that journalists had access to this, and "increasingly started to focus on the party's messy ways of self-organizing and ubiquitous conflicts carried out in public" (Ringel, 2019, p. 7). As a result of this scrutiny, and as a countermeasure to what we might call excess transparency, the party gradually developed an intricate practice of sharing particular parts of their work online, while also developing a range of offline practices to ensure that some discussions did not reach the public eye. With the sociomaterial lens, we see that the digital, the physical, and the social were interwoven in new organizational practices.

Connecting legitimacy and transparency: The case of a pharmaceutical company

We have seen in this chapter how the advent of digital communication platforms accentuates the interconnectedness of legitimacy and transparency. In conclusion, let us return to a case that shows this interconnectedness and demonstrates the organizational work involved in manufacturing legitimacy and transparency. In Chapter 8 on communication and interactivity, we discussed some of the challenges faced by a large pharmaceutical company attempting to reach out to their stakeholders via Twitter. This account is also relevant to discussions of legitimacy and transparency. We saw how this new type of stakeholder engagement was an instance of strategic work with legitimacy. In addition, we saw how the use of Twitter – or social media more generally – could serve as tools to either appear or become more transparent. In practice, however, the company found it difficult to become transparent to a larger audience, since their communication on Twitter was initially only directed to the stakeholders, with whom they already had a close relationship (Castelló et al., 2016). The lesson here is that there is no *a priori* crowd 'out there' waiting to engage with a company's tweets. Transparency has to be manufactured. It does not follow automatically from using new digital platforms. Part of this manufacturing of transparency also has to do with how the digital platforms are used – in this particular organization, the work of appearing legitimate and transparent required new practices and types of communication. The employees responsible for the social media engagement had to be available beyond normal working hours, so that the organization did not appear unresponsive, and employees had to understand how to enter conversations with the community in a way that was perceived as meaningful and, hence, legitimate.

Earlier we argued for the increasing importance of interactivity, and here we see how legitimacy and transparency also seem to be dependent on an understanding of the interactivity embedded in today's digital communication platforms. We also discussed the increasing importance of networked forms of organizing. In relation to legitimacy, Castelló and colleagues (2016) argue that social media have created the conditions for 'networked legitimacy strategies'. This is because they have the potential to transform power relations, challenge hierarchical structures, and encourage two-way communication. But as this case shows, a strategic *potential* does not ensure a successful, transparent stakeholder communication.

The themes of legitimacy and transparency are not new to organizations. What is new is the way digital technologies are interwoven with the construction of legitimacy and transparency. We have treated legitimacy and transparency as interconnected phenomena. We believe that this interconnection is more important now precisely because digital technologies afford new expectations to – and means to produce – transparency, and because new accountability practices also influence legitimacy work. A sociomaterial perspective on legitimacy and transparency in digital organizing offers a way to capture the everyday practices that go into constructing both phenomena, and to discern the role that digital technologies play in affording a particular type of legitimacy.

? QUESTIONS

1. What kind of new legitimacy issues arise as a result of the explosion in information sources on the internet?

2. How should organizations redefine their stakeholder relations and stakeholder management, given how digital communication platforms can be (and are) used?

3. How do specific digital technologies afford different transparency practices?

4. How do specific transparency practices influence particular organizations' legitimacy?

5. Which methods do organizations use to manufacture transparency? Can we speak of some methods being more appropriate than others in a given situation?

References

Barros, M. (2014). Tools of legitimacy: The case of the Petrobras corporate blog. *Organization Studies*, 35(8), 1211–1230.

Berg, O. (2011). 3 reasons why organizations need to increase transparency. *CMS Wire*, July 5. Retrieved from http://www.cmswire.com/cms/enterprise-collaboration/3-reasons-why-organizations-need-to-increase-transparency-011886.php

Bernstein, E. S. (2017). Making transparency transparent: The evolution of observation in management theory. *Academy of Management Annals*, 11(1), 217–266.

boyd, d. & Crawford, K. (2012). Critical questions for big data. *Information, Communication & Society*, 15(5), 662–679.

Businessdictionary.com (2019). Transparency. Retrieved from http://www.businessdictionary.com/definition/transparency.html

Castelló, I., Etter, M. & Årup Nielsen, F. (2016). Strategies of legitimacy through social media: The networked strategy. *Journal of Management Studies*, 53(3), 402–432.

Christensen, L. T. & Cheney, G. (2015). Peering into transparency: Challenging ideals, proxies and organizational practices. *Communication Theory*, 25(1), 70–90.

Clegg, S. R., Rhodes, C. & Kornberger, M. (2007). Desperately seeking legitimacy: Organizational identity and emerging industries. *Organization Studies*, 28(4), 495–513.

Du Plessis, E. M. (2014). Whistleblower-ordninger: Offentlighed, kritik og risikohåndtering. *Slagmark*, 60, 139–154.

Ellerbrook, A. (2010). Empowerment: Analysing technologies of multiple variable visibility. *Surveillance & Society*, 8(2), 1477–7487.

Espeland, W. N. & Sauder, M. (2007). Rankings and reactivity: How public measures recreate social worlds rankings and reactivity. *American Journal of Sociology*, 113(1), 1–40.

Flyverbom, M., Leonardi, P. & Stohl, C. (2016). The management of visibilities in the digital age. *International Journal of Communication*, 10, 98–109.

Fombrun, C. J. & van Riel, C. B. M. (2004). *Fame & fortune: How successful companies build winning reputations*. Upper Saddle River, NJ: FT Prentice Hall.

Freeman, R. E. (1984). *Strategic management: A stakeholder approach*. Cambridge: Cambridge University Press.

Friedman, M. (1962). *Capitalism and freedom*. Chicago, IL: University of Chicago Press.

Friedman, M. (1970). The social responsibility of business is to increase its profits. *The New York Times Magazine*, 13 September. Retrieved from http://umich.edu/~thecore/doc/Friedman.pdf

Grolin, J. (1998). Corporate legitimacy in risk society, the case of Brent Spar. *Business Strategy and the Environment*, 7(4), 213–222.

Gulbrandsen, I. T. & Just, S. N. (2013). Collaboratively constructed contradictory accounts: Online organizational narratives. *Media, Culture & Society*, 35(5), 565–585.

Hansen, H. K. & Flyverbom, M. (2015). The politics of transparency and the calibration of knowledge in the digital age. *Organization*, 22(6), 872–889.

Hartz, R. & Steger, T. (2010). Heroes, villains and 'honourable merchants': Narrative change in the German media discourse on corporate governance. *Organization*, 17(6), 767–785.

Hotten, R. (2015). Volkswagen: The scandal explained. *BBC News*, December 10. Retrieved from http://www.bbc.com/news/business-34324772

KL (2015) Notat – Brugerportalsinitiativet, January 26. Retrieved from https://backend.folkeskolen.dk/-/3/8/offentligtnotatombrugerportalsinitiativet.pdf

Lazer, D., Kennedy, R., King, G. & Vespignani, A. (2014). Google Flu Trends still appears sick: An evaluation of the 2013–2014 flu season. Retrieved from https://gking.harvard.edu/publications/google-flu-trends-still-appears-sick%C2%A0-evaluation-2013%E2%80%902014-flu-season

Lombardi, K. S. (1997). New surveillance: Day-care cyber visits. *The New York Times*, March 16. Retrieved from https://www.nytimes.com/1997/03/16/nyregion/new-surveillance-day-care-cyber-visits.html

Meyer, J. W. & Rowan, B. (1977). Institutionalized organizations: Formal structure as myth and ceremony. *The American Journal of Sociology*, 83(2), 340–363.

Mitchell, R. K., Agle, B. R., & Wood, D. J. (1997). Toward a theory of stakeholder identification and salience: Defining the principle of who and what really counts. *Academy of Management Review*, 22(4), 853–886.

Orlikowski, W. J. & Scott, S. V. (2014). What happens when evaluation goes online? Exploring apparatuses of valuation in the travel sector. *Organization Science*, 25(3), 868–891.

Perrow, C. (1991). A society of organizations. *Theory and Society*, 20(6), 725–762.

Perry, N. (1998). Indecent exposures: Theorizing whistleblowing. *Organization Studies*, 19(2), 235–257.

Porter, M. & Kramer, M. (2011). Creating shared value. *Harvard Business Review*, 89(1/2), 62–77.

Rawlins, B. (2008). Give the emperor a mirror: Toward developing a stakeholder measurement of organizational transparency. *Journal of Public Relations Research*, 21(1), 71–99.

Ringel, L. (2019). Unpacking the transparency-secrecy nexus: Frontstage and backstage behaviour in a political party. *Organization Studies*, 40(5), 705–723.

Scott, C. R. (2013). *Anonymous agencies, backstreet businesses, and covert collectives: Rethinking organizations in the 21st century*. Stanford, CA: Stanford University Press.

Sifry, M. L. (2011). *Wikileaks and the age of transparency*. New York: OR Books.

Suchman, M. C. (1995). Managing legitimacy: Strategic and institutional approaches. *The Academy of Management Review*, 20(3), 571–610.

Transparency.org (2019). Our Organisation: Mission, Vision and Values. https://www.transparency.org/whoweare/organisation/mission_vision_and_values/0/

Vaara, E., Tienari, J. & Laurila, J. (2006). Pulp and paper fiction: On the discursive legitimation of global industrial restructuring. *Organization Studies*, 27(6), 789–813.

10 POWER AND EMPOWERMENT

> In this chapter, we show how the techno-optimist literature on digital organizing has treated power as a uniform concept. Drawing on organization scholars Peter Fleming and André Spicer (2014), we offer an alternative view of power in organizations based on four dimensions: (1) coercion: the ability to force others to change behavior, (2) manipulation: the ability to prevent certain things from happening, (3) domination: the ability to influence the preferences of others, and (4) subjectification: the process by which individuals are transformed into subjects. Our main argument is that while early studies of digital organizing acknowledged the multidimensional character of power, the techno-optimist literature comes with more commonsense understandings of power. In fact, many of the techno-optimist accounts of digital organizing view power simply as the ability to 'get things done' and empowerment is spoken of in unambiguously positive terms. If power is viewed as a repressive force that prevents people from pursuing their real interest, empowerment is seen as a force of liberation. And if digital technology is viewed as the primary driver of individual autonomy, digitalization is easily equated with widespread empowerment of ordinary people. However, as we show in this chapter, empowerment is not necessarily an antonym to power. Inspired by Michel Foucault we argue that empowerment can serve as a different form of power-wielding that works through the 'hearts and minds' of people rather than through simple coercion. Such an understanding of power allows us to understand aspects of digital organizing which are perhaps counter-intuitive.

Why power and empowerment?

A frequently used buzzword in contemporary business lingo is that of 'employee empowerment'. In the academic literature, employee empowerment typically refers to the process of granting employees control over their own work by providing them with a certain degree of autonomy on the job (Honold, 1997). With the rise of new digital platforms for collaboration (see Chapter 6), the notion of employee empowerment has reached a whole new level. Instead of having to negotiate the limits to autonomy within the boundaries of an ordinary workplace, workers can now break free from the constraints of traditional labor processes and become online freelancers or even 'digital nomads', traveling the world while selling their services on online platforms (Brett, 2015). In this way, people can work *whenever* they want, *wherever* they want, and charge *whatever* they want, as long as they are properly connected to the World Wide Web. Sally-Anne Barnes (2013, p. 27) put it in a report for the European Commission, these new digital platforms "can be viewed as empowering individuals by creating and providing a space in which they can self-select work, be creative and/or interact to solve problems as part of a wider community".

Every year, an increasing number of people make their living as freelancers. A recent study found that 36 percent of the American workforce earned freelance income in 2017 alone, and this proportion is expected to exceed 50 percent by 2027 (Edelman Intelligence, 2017). The study also showed that the expansion of the freelance economy is heavily driven by digital technology. In fact, 71 percent of the study's respondents stated that the amount of work they obtained online has increased during the past year. When asked about the benefits of this type of work, and why they chose to engage with online freelancing in the first place, most respondents mentioned 'freedom' and 'flexibility' as their main motivations, which, as we shall see, are two central elements in what we refer to as the discourse of empowerment.

(Em)powering oneself: The case of UpWork

One of the most popular freelance platforms in the world is a California-based platform called *UpWork* (previously *oDesk*). The platform is currently home to more than 12 million registered freelancers, and it curates more than three million job postings each year. Through the UpWork platform, freelancers are able to display their skills in a professional manner and respond to job postings with only a few clicks. All one has to do, in order to access this world of job opportunities, is to create a personal profile and start looking for work. Furthermore, when creating a personal profile, the UpWork platform offers useful advice for the inexperienced freelancer. For instance, it recommends that users upload a personal photo to the website, since "freelancers with a friendly, professional-looking portrait are hired 5 times more often than those without a photo" (UpWork website, 2017, n.p.). And if one is in doubt as to what constitutes a 'professional-looking' photo, the platform also contains a how-to guide that, step by step, helps users navigate the world of business portraits:

> You have excellent experience and top-notch skills, but you're having a hard time landing a job. Sound familiar? If so, consider the first impression your profile makes with prospective clients ... Clients find smiling freelancers more warm, friendly, and trustworthy. Not used to smiling for the camera? Try pretending that you are greeting your best friend. (UpWork blog, 2017, n.p.)

UpWork is far from alone in dispensing such advice. Thousands of websites and YouTube videos offer advice on how to become successful in the world

POWER AND EMPOWERMENT

Image 10.1 Freelance platforms are new sites for trading one's labor. To some, freelancing offers the flexibility of working from home, from an exotic beach, or while sitting in a café. To others, freelance platforms are a stressful marketplace with cutthroat competition, no rest time, and with no guarantee of a minimum income.

Source: FXQadro, iStock.

of online freelancing. The advice differs, but certain types of recommendations recur. For instance, the majority of sites claim that the successful freelancer should be industrious, self-confident, competitive, systematic, friendly, and trustworthy. The key to success is thus to create a personal 'brand' that encapsulates all of the above. As one self-help website puts it: "It doesn't matter if you're just one person offering one particular service, you still need to treat yourself as a professional business selling something premium – and in return, people will treat you as such" (Socialmediatoday, 2017, n.p.) (Image 10.1).

This freelance version of employee empowerment offers us a very commonsense understanding of power: freelancers are empowered because no one is forcing them to do anything. Freelancers are their own masters, to the extent that they can decide for themselves when and how to work. But that is only until bills have to be paid.[1] When that happens, it is no longer enough to be a freelancer; one has to become a *successful* freelancer. And

---

1 A 2016 survey showed that freelancers in the creative industries lose £5,394 each year through working for free in order to obtain sufficient experience and exposure. The freelancers participating in the survey had an average of seven years' experience in their respective fields (Chandler, 2017).

as we have seen, this requires the otherwise so autonomous freelancer to be industrious and systematic, competitive and trustworthy, self-confident and friendly, while greeting potential clients in a warm and welcoming manner.

The understanding of power that fuels the discourse of empowerment, as well as the notion of online freelancing with its message of liberation from control by others, is clearly one that views power as something that is done to someone. In this view, the conventional employer is powerful because he or she can make the employee do something that the employee would not have done otherwise. People become freelancers in order to escape these kinds of constraints. But as we shall see in this chapter, power also operates in other ways. More nuanced understandings tend to view power structures as not only coercive, but also more or less self-imposed. For instance, in the example above, we see how online freelancers are encouraged to discipline themselves by constructing a personal brand that projects an image of professionalism and trustworthiness, compelling them to work hard and to greet potential clients (i.e. employers) as if they were 'best friends'. No one is forcing the freelancers to do this. The compulsion must come from within. To capture such power dynamics, the chapter argues for developing an analytical sensitivity toward how power and empowerment operate in digital organizing.

Power as a theme in organization studies

Within organization studies, power seems to be an almost omnipresent phenomenon. In fact, it can be difficult to find books and articles about organizational dynamics that do not address questions of power in some form. As Stewart Clegg and colleagues (2006, p. 3) put it: "power is to organization as oxygen is to breathing". At the same time, however, power is also an "essentially contested concept" (Gallie, 1955, p. 169), which means that there is little consensus as to what the term implies. As we saw above, some authors employ a somewhat intuitive understanding of power closely related to coercion (see Dahl, 1957), while others view power as the ability to make a difference in the constitution of identities (see Foucault, 1982). These opposing views of power may be seen as different conceptualizations of power *as such*, but they may also be viewed as different dimensions or 'faces' of power in and around organizations (Fleming & Spicer, 2014).

Power as coercion

Reviewing the vast literature on power within organization studies, as well as beyond, one discovers that there are roughly four ways of responding to the question "What is power?" The first approach is closely associated with the work of political scientist Robert Dahl (1957, 1961) who, in building on the work of Max Weber, advanced the compelling argument that power should be measured by the extent to which A is able to make B do something that he or she would not have done otherwise. As Dahl (1957, p. 203) puts it himself: "The base of an actor's power consists of the resources – the opportunities, acts, objects, etc. – that he [*sic!*] can exploit in order to effect the behavior of another". Fleming and Spicer (2014, p. 241) refer to this face of power as 'coercion', since it operates through the "direct mobilization of power". As we saw above, this is the understanding of power that guides the discourse of empowerment, as well as the idea that online freelancing is the expression of enhanced freedom and flexibility. Since power is viewed as something that can be possessed (e.g. by employers), it makes perfect sense to think of it as something one can escape from by becoming self-employed.

In organization studies, this understanding of power as coercion has been used to study a wide variety of phenomena, ranging from resource allocation to decision-making processes. The common denominator across these studies is that they share a focus on groups and individuals as the primary loci of power. One example is Robert Michels' (1911/1962) famous study of political parties and trade unions in twentieth-century Europe. Through first-hand experience within the German Social Democratic Party in particular, Michels observed how otherwise highly democratic and egalitarian parties became authoritarian and oligarchic as soon as they entered the formal political system. The explanation for this oligarchic drift was that the political leadership often developed a desire to maintain the party's status and position in the system, which made it react "with all the authority at its disposal against revolutionary currents which exist within its own organization" (Michels, 1911/1962, p. 371).

Another example is Crozier's (1964) ethnographic study of two French bureaucracies, in which he argued that power resides in the individual employee's ability to reduce complexity for others. For instance, in one bureaucracy (a monopolistic manufacturer of cigarettes and matches), Crozier observed how the maintenance engineers were surprisingly powerful because they controlled the firm's only real source of uncertainty: the production system. What made the engineers powerful was their ability to make others dependent on their discretionary actions – if they did not succeed, no one succeeded. The power of the engineers ultimately reversed the formal chain of command, forcing managers and directors to ask the engineers for advice on how to run the organization. In both examples, a group of people (the political leadership or the maintenance engineers) is seen as powerful because they visibly *possess* the ability to alter the behavior of others.

Power as manipulation

The second approach to the study of power was initially advanced by Peter Bachrach and Morton Baratz (1962, 1963) as a response to Dahl's coercion-focused understanding. While the first face of power pertains to moments where power is mobilized in an attempt to directly impact the behavior of others, Bachrach and Baratz saw a more subtle dimension of power, overlooked by Dahl, which they called 'non-decision-making'. What they meant by this was that power not only resides in the ability to make something happen, but also in the ability to ensure that something *does not* happen. "Of course power is exercised when A participates in the making of decisions that affect B", they argued. "But power is also exercised when A devotes his [*sic!*] energies to creating or reinforcing social and political values and institutional practices that limit the scope of political processes to … those issues which are comparatively innocuous to A" (Bachrach & Baratz, 1962, p. 948). In doing so, A ensures that his or her preferences are satisfied without actually mobilizing power in any direct sense. Fleming and Spicer (2014) refer to this face of power as 'manipulation', since it is constituted by an implicit or hidden form of agenda-setting.

This view of power has not been as widely deployed within organization studies as the coercive definition, which may have to do with the fact that it can be rather difficult to identify things that do not happen. As Clegg (1989) notes, how does one know whether someone actually *intended* to manipulate the scope of decision-making, or whether certain events just unfolded more or less incidentally? Nonetheless, the second, non-decisional face of power has been used by organizational scholars to explore how organizational processes are manipulated in relation to issues of gender-bias and collective bargaining, among other issues. Here, the common argument is that the ability to manipulate decision-making processes hinges on one's position in different social networks. One example is Rosabeth Kanter's (1977) study of 'men and women of the corporation', which is concerned with structures of advancement in large hierarchical organizations. At the time of the study, only 2 percent of all top managers were female (Puffer, 2004), which made critical researchers wonder whether certain organizational configurations constituted a 'glass ceiling' that prevented women from achieving upward mobility. Based on observations in a large manufacturing firm, Kanter argued that managerial advancement is often determined not by merit or seniority, but by people's ability to navigate the "shadow political structure" of the organization – a structure that implicitly prevented female employees from gaining access to the upper echelons by impeding them from forging alliances with resourceful actors (Kanter, 1977, p. 197). Hence, instead of asking *what* makes people powerful, as the coercion-focused studies often do, the manipulation-focused studies usually explore *how* hidden structures privilege some and disenfranchise others.

Power as domination

Despite their differences in focus, representatives of the two first types of power-wielding agree that power is something to be possessed and mobilized episodically, depending on people's access to scarce resource as well as their position in social networks. The third face of power extends this so-called 'behavioral' view by adding an ideological dimension to the question of power. Steven Lukes (1974) first advanced the notion of a third face of power in his widely cited book, *Power: A radical view*. Here, Lukes argued that the most effective type of power is not one that can be possessed in any meaningful way, but instead one that operates through the subtle shaping of interests and preferences. In other words, the most 'powerful' type of power is one that influences people without their knowing – and sometimes even in a manner inconsistent with their 'real' interests. While this understanding of power naturally raises some thorny questions about the distinction between 'objective' and 'subjective' interests (see Benton, 1981), the ambition is to explore how and why certain segments of society quietly accept or even reinforce conditions that seem far removed from their own personal well-being. For instance, why do people who rely on government subsidies for survival actively support welfare cutbacks? In many cases, the answer to this 'great paradox' (Hochschild, 2016) pertains to the diffusion of *dominant* ideologies in contemporary society, which is why Fleming and Spicer (2014) refer to this third face of power as 'domination'.

Quite a few organization studies base their theorizing on this understanding of power, particularly within the field of research known as 'institutional theory'. One example is John Meyer and Brian Rowan's (1977) seminal work on institutions as 'rationalized myths' that organizations must comply with in order to attain legitimacy (see also Chapter 9). The bureaucratic form is an example of such a rationalized myth. The reason why "modern societies are filled with rationalized bureaucracies", the authors suggested, is not necessarily because the bureaucratic form is superior in terms of effectiveness, but because it is the most legitimate organizational form (Meyer & Rowan, 1977, p. 345). As such, the bureaucratic form has become an almost inescapable point of reference in the ideology of modernization, obscuring the ability of some organizations to pursue their 'real' interests, in case these are incompatible with the bureaucratic form. This perspective has been further advanced by Paul DiMaggio and Walter Powell (1983) who suggested the notion of 'institutional isomorphism' as a way of understanding why organizations often imitate each other instead of seeking out unique modes of organizing, despite the fact that this might have provided them with a competitive advantage. As the authors note: "Organizations compete not just for resources and customers, but for political power and institutional legitimacy, for social as well as economic fitness" (DiMaggio & Powell, 1983, p. 150). It is this constant quest for social fitness that, for organizations and individuals alike, undergirds the domineering dimension of power.

Power as subjectification

The first three faces all rely on the same basic assumption: namely that power is something that is done to someone at certain points in time. Power is present when A significantly affects B, whether or not this takes the form of coercion, manipulation, or ideological domination. Two important implications follow from this basic assumption: (1) that power is *derivative*, in the sense that it flows from a particular source such as an actor or a structural arrangement, and (2) that power is *escapable*, in the sense that power-free spaces exist (Dyrberg, 1997). The fourth face of power, manifested in what Lukes (2005, p. 12) calls the 'ultra-radicalism' of Michel Foucault, contains a dramatic break with both of these. Instead of viewing power as a repressive force that somehow prohibits individuals and organizations from pursuing their real interests, Foucault (1977, 1982) conceived of power as a *productive* force that underlies the constitution of knowledge in general. Power does not obscure what is 'true' or 'real' – it defines what counts as such. In other words, power and knowledge are two sides of the same coin. This view of power is perhaps most clearly illustrated in relation to questions of identity. As Foucault notes: "It's my hypothesis that the individual is not a pre-given entity which is seized on by the exercise of power. The individual, with his identity and characteristics, is the product of a relation of power exercised over bodies, multiplicities, movements, desires, forces" (Foucault, 1980, pp. 73–74).

For Foucault, what we call 'identity' – knowledge of who we are as individuals and organizations – is the provisional result of ongoing struggles between competing discourses. By discourses, Foucault refers to systems of utterances, which "causes words and things" to "hold together" in a certain way (Foucault, 1966, p. xix). Discourses shape how we understand the world, including ourselves as part of that world. As such, identity does not exist prior to written, verbal, or material 'utterances'. Identity emerges as a product of these (hence, the notion of power as 'productive'). Foucault was particularly interested in the process by which individuals are transformed into subjects; that is, the process by which people are encouraged to recognize themselves as a particular type of individual (Foucault, 1982). For instance, in his later writings, Foucault (1978) investigated how people in the eighteenth and nineteenth centuries became increasingly tied to their own sexual orientation, and how the production of scientific knowledge about sexuality at that time helped foster subjects who were more conscious of their own pleasures and desires. To capture the dimension of power that underlies the production of subjects, Fleming and Spicer (2014) refer to this fourth face of power as 'subjectification'.

During the past three decades, the Foucaultian understanding of power has gained widespread recognition within organization studies. To give an example, a topic in studies that adopt this view could be what Mats Alvesson and Hugh Willmott (2002) call 'identity regulation' – that is, the alignment of employee subjectivity with corporate interests. One classic example is Gideon Kunda (1992) study of corporate culture in a Silicon Valley company pseudonymously referred to as 'Tech'. What characterizes Tech's corporate culture is an almost complete absence of rules and regulations. In fact, there is only one rule that employees must follow at all times: *Do What's Right!* This lack of rules and formal structure naturally provides the people working at Tech with a significant degree of autonomy on the job, but it also deprives them of the ability to know when their job is done. This type of 'designed ambiguity' thus installs a particular sense of self in the employees, which animates them to labor tirelessly in an effort to satisfy what amounts to fundamentally insatiable demands. The employees labor willingly, however, because they take pride in their work and see their employment as a privilege. As such, the power that permeates organizations like Tech is not captured by any traditional understanding of the word, since power does not flow causally from A to B, but works through the 'hearts and minds' of those being subjectified. As a senior manager, who continuously refers to Tech's culture as a religion, puts it: "Power plays don't work. You can't *make* 'em [the employees] do *any*thing. They have to want to. So you have to work through the culture. The idea is to educate people without them knowing it. Have the religion and not know how they ever got it!" (Kunda, 1992, p. 5).

This quote not only summarizes how culture functions as a control mechanism in Tech, but it also indicates that the senior manager fails to see the type of 'power plays' that underpin his own managerial practices. Since the early 1990s, a plethora of studies on subjectification in organizations have been published in relation to an extensive list of organizational phenomena (e.g. Knights & Morgan, 1991; Townley, 1993; Brewis et al., 1997; Svenningson & Alvesson, 2003; Bergström & Knights, 2006; Laine & Varaa, 2007; Costas & Kärreman, 2013; Husted, 2018). The common denominator across these studies is an awareness of the fact that organizations often spend as much energy 'producing the appropriate individual' (Alvesson & Willmott, 2002) as they do producing goods and services – parallel to the introductory description of online freelancers who must transform themselves into particular types of individuals in order to succeed. As we shall see in the following sections, this awareness was indeed present in some of the earliest studies of digital organizing, but it slipped into the background with what we call the 'techno-optimist' literature.

Power in the age of the 'smart machine'

One of the earliest attempts to understand power in organizations that coordinate their activities through the use of digital technology is Shoshana Zuboff's (1988) influential monograph, *In the Age of the Smart Machine*. The main argument of her book is that new computer-based technology anno 1988 has two possible functions. First, it automates work processes by turning manual labor into machine labor, adding a previously unprecedented level of speed and precision. However, this is not a novel feature of the 'smart machine', but a common theme in most workplace technology. For instance, the automation of work carried out by the assembly line could be seen as an attempt to add precision and speed to the manufacturing process. What makes computer-based technology (or what we call 'digital technology') unique is that it not only *automates* but also *informates* by registering data about the automated activities. Computers not only accomplish certain tasks; they also register when, how, and why these tasks were completed.

In the latter part of her book, Zuboff uses the duality of automation and information to describe the implications of computer-based technology on power relations at work. She begins by considering the consequences for coercive power (she uses the term 'imperative control'), which is a mode of power often manifested in hierarchical relations of command and obedience. In accordance with scholars of workplace discipline before her (e.g. Braverman, 1974), Zuboff argues that the automating aspect of technology tends to accentuate managerial hierarchies because it deskills workers and deprives them of the right to structure their own work. However, in the case of complex digital technology, ambiguity is injected into the heart of the matter. Suddenly, managers find themselves in situations where *they* do not understand how the technology works, and where workers may be more skilled at operating the machines or accessing sensitive information. This *reskilling* of workers constitutes an immense challenge to management as an occupation (see also Burawoy, 1979). After all, why do we need centralized authority when the 'smart machine' does a large portion of the thinking? And what is the role of coercive power when it is the workers who master the machine?

From a managerial point of view, the main problem with digital technology is not only that managers do not understand how the technology works, but, as Zuboff (1988, p. 308) puts it: "The more blurred the distinction between what workers know and what managers know, the more fragile and pointless any traditional relationships of domination and subordination between them become". As such, the informating aspect of digital technology has the potential to subvert the traditional logic of managerialism, because it places knowledge and thus power in the hands of workers. This new constellation forces managers to find new ways of overseeing their employees. Ironically, what makes digital technology so subversive (an excess of data) also provides the grounds for a new type of power that Zuboff refers to as 'panoptic power', the power of surveillance.

Discipline and informate: The case of tax administration

With the notion of panoptic power, we arrive where we left off in the previous section, namely at the pen of French historian Michel Foucault. Panopticon (which literally means 'all-seeing') is the name of a rather sophisticated architectural structure, developed by the brothers Jeremy and Samuel Bentham in the late eighteenth century, as a response to the practical problems associated with housing and monitoring large numbers of prison inmates. The main components of the panopticon are a circular structure, inhabited

Image 10.2 A Cuban prison built as a panopticon.

Source: Friman, Wikimedia.

by the inmates, and a watchtower with blacked-out windows in the middle, occasionally occupied by guards. The point of the panopticon is to enable a small number of guards to supervise a large number of inmates, which is made possible by an architectural design that allows the guards to see all inmates from one single point of observation. However, the terrifying genius of the panopticon is not so much that it allows the guards to be 'all-seeing', but that it installs a mode of self-inflicted discipline in the inmates. Because of the blacked-out windows, the inmates can never be certain that they are actually being watched. All they know for sure is that they are *potentially* being watched. So, while the inmates in the panopticon enjoy quite a bit of freedom, in the sense that they are not being directly controlled, they quickly end up controlling themselves out of fear from being watched (Image 10.2).

Even though Jeremy Bentham (1791, p. 3) initially framed his panopticon as "a new mode of obtaining power of mind over mind", it was not until Foucault's (1977) groundbreaking book *Discipline and Punish* that the notion of the panopticon gained influence in the power literature. According to Foucault, the panopticon was more than an architectural design; it was also a metaphor for a wider shift in the exercise of power within the Western penal system. While in earlier times, judicial punishment typically targeted the criminal body (by way of torture or execution), the Enlightenment era ushered in a shift toward a more 'humane' form of punishment targeting the criminal mind. As such, during the nineteenth century, physical punishment was replaced or supplemented by mental discipline. Discipline thus became the favored mode of power within the penal system, but also within society more generally. Corporeal punishment was gradually abolished in schools, families, and factories and replaced by more sophisticated disciplinary techniques. The essence of these disciplinary techniques is manifested in the architecture of the panopticon – an arrangement where no one is directly punished, but where everyone is thoroughly disciplined by the threat of constant surveillance.

Zuboff deploys the metaphor of the panopticon to describe the mode of power that is enabled by the informating aspect of digital technology. Her point is that the excess of data produced by these technologies render possible a situation where management is capable of surveilling employees in a way that they could not before. Prior to the introduction of the 'smart machine', management had to be physically present in order to monitor the performance of workers and to understand the nature of the work processes. Today, however, digital technology has enabled management to supervise and discipline subordinates 'at a distance' (Jones, 2000). Control at a distance has thus given birth to a disciplinary workplace more omnipresent than anyone could have dreamt of. As Zuboff bluntly notes: "Information systems that translate, record, and display human behavior can provide the computer age version of universal transparency with a degree of illumination that would have exceeded even Bentham's most outlandish fantasies" (Zuboff, 1988, p. 322).

Present-day digital technologies support management in new ways by rendering workflows visible and generating management information in real time. These technologies allow managers to disregard all instances where work proceeds according to the established standard and to flag any deviations from the norm. For instance, in a section of the Danish tax administration, case-handling processes are registered and depicted as a 'river', running in a particular way. The idea is that when managers look at this data-driven visualization, they can see where irregular 'streams' break free. The streams represent instances where a tax case handler opens the wrong software, lingers for too long on a given screen, or deviates in other ways from the standard pace or trajectory of work. The mechanisms behind these visualizations are hidden from both employees and managers, but offer new tools to monitor, supervise, and control employees.[2]

Naturally, this does not mean that digital technology is the computer-based equivalent of Bentham's sinister penitentiary structure – a technological dystopia of Orwellian proportions. It means, however, that the vast amount of data produced by such technology provides a fertile ground for management to assess employee performance. It is thus a tool for 'producing the appropriate individual' (Alvesson & Wilmott, 2002). Zuboff's more general point is that within digital organizations, different

---

[2] This example is based on an interview with an employee in the Danish tax administration and it is part of Ursula Plesner and Lise Justesen's ongoing research on the digitalization of public sector organizations.

Table 10.1 Functions of digital technologies in the workplace.

Aspect of digital technology	Organizational function	Implications for power
Automating	Turns manual labor into computer labor, while adding high levels of precision, speed, and repetition	Concentrates coercive power in the hands of managers by removing 'brain work' from the shop floor
Informating	Produces data about automated tasks, registering how, when, and why certain tasks were completed	Subverts coercive power relations, but provides conditions for disciplinary forms of power

dimensions or 'faces' of power are operating in tandem. While the automating aspect of digital technology enhances authoritarian or coercive forms of power by intensifying hierarchical relations of command and obedience, the informating aspect of digital technology simultaneously subverts authority *and* establishes the basis for more disciplinary forms of power. To put it differently, subjectification takes over where coercion fails (Table 10.1). Hence, Zuboff employs a multidimensional understanding of the concept of power that allows her to see it as an omnipresent phenomenon, which is not only non-derivative but also non-escapable (for more studies that apply similar perspectives to issues of digital organizing, see Ball & Wilson, 2000; Brooke, 2009; Coombs et al., 1992; Doolin, 1998; Grint & Woolgar, 1997; Knights & Willmott, 1988; Scarbrough & Corbett, 1992; Sewell & Wilkinson, 1992). In the following section, we shall see how this multidimensional understanding slips into the background in the techno-optimist part of the organization literature that focuses almost exclusively on empowerment in digital organizing.

Digital empowerment

The concept of empowerment is usually traced back to the Brazilian educator and Marxist thinker Paulo Freire (1970), who used the term to highlight the emancipatory potential of critical pedagogy in oppressive societies (Fortunati, 2014). Through education, Freire argued, the oppressed would begin to see their oppression as an object of reflection and automatically begin the process of organizing their struggle for liberation. Recalling our previous discussion of the four faces of power, it is not difficult to decipher the basic assumptions behind this initial concept of empowerment. Power is exercised by powerful oppressors (A) over the powerless oppressed (B), although the oppressed have the potential to escape the grip of power through the empowering force of education. This conceptual legacy is important to keep in mind when considering how the notion of empowerment has been used in the techno-optimist literature on digital organizing. Here, the emancipatory power of education is, in a sense, replaced by digital technology as the all-empowering force.

One of the most emblematic examples of this digital empowerment literature is Clay Shirky's (2008) international bestseller, *Here Comes Everybody: The power of organizing without organizations*.[3] The book opens with an anecdote about an American woman who leaves her cell phone in an NYC taxi, only to see it picked up by a poor girl from Queens who stubbornly refuses to return the phone to its rightful owner. However, upon mobilizing thousands of people through a designated website detailing the events surrounding the theft, the woman eventually manages to get her phone back, and the girl from Queens is arrested – all of this despite the police's reluctance to take action. According to Shirky, this story has many themes, but one of them concerns "the power of group action, given the right tools" (Shirky, 2008, p. 7). Thanks to the internet, the woman and her tech-savvy friends were able to force the otherwise so powerful bureaucracy of the New York Police Department (NYPD) into action, thus restoring a sense of justice to the many people following the events online.

Despite the central status of power in the book's subtitle, the word is never treated in any substantial matter throughout the text. It is simply assumed that power is a term that requires no explanation. This shortcoming is also visible in several other examples of the techno-optimist literature on digital organizing such as *Crowdsourcing: How the power of the crowd is driving the future of business* (Howe, 2008), *The Art of Social Media: Power tips for power users* (Kawasaki & Fitzpatrick, 2014), or *Platform Strategy: How to unlock the power of communities and networks to grow business* (Reillier & Reillier, 2017). Nowhere in these books is the concept of power spelled out. Instead, they offer an abundance of memorable anecdotes like the one above that implicitly gives the reader a sense of what power might mean to these authors. In some cases, power is conceived as the capacity of ordinary people to make otherwise powerful actors alter their behavior (this is the understanding that Shirky subscribes to), but in most cases, power is simply understood as "the ability to get things done, to mobilize resources", as Kanter (1977, p. 166) puts it. What makes digital technology a force of empowerment is thus the fact that more people today are able to 'get things done' than before digitalization.

---

3   Perhaps in recognition of his over-optimism about digitalization, Shirky toned down the more sweeping subtitle for the second edition of the book. It now reads, *How change happens when people come together.*

From coercion to subjectification: The case of HEROes

Power in the techno-optimist literature is thus seen as something to be possessed depending on one's access to scarce resources and one's position in social networks. From this perspective, it makes perfect sense to speak of empowerment – the process of *transferring* power from the powerful to the powerless – in unambiguously positive terms. This way of thinking about power is perhaps illustrated most clearly in Josh Bernoff and Ted Schadler's (2010a) book, *Empowered: Unleash your employees, energize your customers, and transform business*. The main premise of their book is that digital technology (mobile technology in particular) has provided customers with unprecedented tools for inflicting 'lasting brand damage' on multinational corporations. By simply posting an angry tweet or composing a critical YouTube video, present-day customers are able to challenge the image of previously untouchable megabrands, which is why corporations urgently need to consider how to respond. Fortunately, the authors are ready with an answer: employee empowerment. In order to match the angry customers, employees must be given the freedom and tools to fight back in this digital contest between customers and businesses. This is the only way, it seems, "to keep your company safe as well as responsive" (Bernoff & Schadler, 2010b, p. 101).

However, Bernoff and Schadler do not declare that all employees should be equally empowered. Significantly more freedom should seemingly be awarded to what they call the HEROes (an abbreviation for 'highly empowered and resourceful operatives') of the corporations, who are creative employees who innovate with technology for the benefit of business. According to the authors, the best corporate response to digitization is thus to "encourage HERO-driven innovation without generating chaos". This strategy is not particularly easy, given the fact than only around 20 percent of all information workers should be considered HEROes (Bernoff & Schadler, 2010b, pp. 97–98). Hence, in this line of thinking, empowerment is all about motivating employees to operate in line with those previously known as the ones 'in power' (i.e. management). An example of this empowerment strategy can be seen in the American tech retailer Best Buy, which has successfully developed a HERO-driven innovation called Twelpforce. Twelpforce is an application that allows Best Buy employees to respond to customer complaints aired publicly on Twitter. In this way, the company's employees can swiftly respond to criticisms – from home or "over a weekend" – because they no longer depend on the spatiotemporal constraints of traditional workplaces. Twelpforce only exists, the authors argue, "because Best Buy empowers its employees to come up with technological solutions". For instance, one of these empowered individuals spent "a week of evenings" figuring out how to tap into Google's cloud computing service, while another "took charge of the project and triumphed over legal obstacles including labor laws" (Bernoff & Schadler, 2010b, p. 96). Without this kind of employee empowerment, the authors argue, Twelpforce would only be an unattainable idea inside some manager's head – or, rather, it would be completely inconceivable.

Empowerment and counter-power

As the examples above illustrate, the discourse of empowerment in digital organizing relies heavily on what we have called a commonsense understanding of power, where power is seen as the ability of groups or individuals to 'get things done', often by forcing other people to change their behavior (i.e. power as coercion). However, if we step back for a second and consider the conceptual history of empowerment (Freire, 1970), we find a more sophisticated understanding of power, which is also embedded in the work of another influential writer on digital technology, namely Manuel Castells. Even though Castells has little to say about organization as such, it is nonetheless relevant to consider his distinction between power and 'counter-power', as it helps us understand what a more refined conception of empowerment might look like. In one of his recent books, *Communication Power*, he writes: "Power is the relational capacity that enables a social actor to influence asymmetrically the decisions of other social actor(s) in ways that favor the empowered actor's will, interests, and values … Power relationships are framed by domination, which is the power that is embedded in the institutions of society" (Castells, 2009, p. 10).

What we get with Castells is an understanding of power, which is multidimensional (he calls it 'eclectic'), in the sense that power may assume different forms. Power may be coercive or manipulative, but it may also be embedded in societal institutions that privilege certain interests and suppress others (i.e. power as domination). This broader conception of power allows us to understand how the version of employee empowerment, promoted by authors like Bernoff and Schadler, is a far cry from what critical theorists like Castells would call the 'real' interests of workers. Only corporate interests are being served by having employees work off-hours to develop a software application that then forces them to work even more hours in order to curb criticism from dissatisfied customers. It is business, not labor, which gains from this type of employee empowerment. What makes the Best Buy employees work so hard *against* their own 'real' interests, Castells would argue, is the ideological power embedded in normative ideals like 'having a career' and 'being promoted' or perhaps in notions of 'self-optimization' and 'productivity' (Cederström & Spicer, 2017).

The way for workers to resist such ideological forms of power would be to exercise what Castells (2007, p. 248) calls 'counter-power', which he conceives as "the capacity by social actors to challenge and eventually change power relations institutionalized in society". One of the most effective ways of doing so is by appropriating internet-based technology or, in his words, "technologies of freedom" (Castells, 2009, p. 414). Elaborating on this point, Castells proposes what seems like an almost natural law of empowerment: "The more an individual has a project of autonomy (personal, professional, socio-political, communicative), the more she uses the Internet. And in a time sequence, the more he/she uses the Internet, the more autonomous she becomes" (Castells, 2007, p. 249). Transferring these rather techno-optimist observations to the case of Best Buy, we see that the employees did indeed confirm the first part of Castells' proposition: they had a professional project of autonomy, so they made use of the internet. However, the latter part of the proposition cannot be confirmed, since the employees did not use the internet to "change power relations institutionalized in society". Instead, they chose to reinforce their own subjugation, precisely at the point where they had the tools necessary to exert a counter power.

Counter-power in practice: The case of #MeToo

Still, there are plenty of examples of digital empowerment that are more in line with the Castellsian understanding of counter-power. Recent examples include the widespread social uprisings known as the Arab Spring, which occurred throughout the Middle East and North Africa from 2010 to 2012, leading to massive societal changes in several countries.[4] Although these uprisings were not directly caused by digital technology, they nonetheless benefited hugely from social media platforms like Facebook or Twitter, which allowed them to "survive, coordinate, deliberate and expand" in an unprecedented manner (Castells, 2012, p. 229). A similar example is the #MeToo movement, which saw women across the world unite against sexual harassment, first in the film industry, then well beyond. By exposing their male perpetrators to public outrage through various media outlets, these women managed to get half the world questioning the 'toxic culture' of sexual misconduct that exists in many male-dominated industries (Zacharek et al., 2017). In both cases, digital technology thus aided ordinary people in their struggle for autonomy within a context of oppression (authoritarianism and patriarchy, respectively).

Though most people would agree that the Arab Spring and the #MeToo movement were positive events that helped ordinary people challenge oppressive institutions and expand their autonomy, there are some analytical dangers in viewing them solely as instances of empowerment. Recalling the Foucaultian notion of power as an omnipresent force, we realize that Castells' conception of counter-power cannot account for instances where oppression is problematized as self-inflicted. For instance, in the wake of the #MeToo movement, 100 French women published an open letter in *Le Monde* claiming that the campaign "chains women to the status of eternal

---

[4] Many of these popular uprisings were given names according to the social media platform most frequently used by protesters. While some uprisings were known as 'Twitter Revolutions' or 'Facebook Revolutions', the Tunisian uprising was referred to as a 'WikiLeaks Revolution', because documents published by WikiLeaks helped spawn the wave of protests that ultimately led to the fall of President Ben Ali (Dickinson, 2011).

victims" by framing them as "poor little things dominated by demon phallocrats" (Andrews et al., 2018, n.p.).

Such statements trigger questions like: did the #MeToo campaign serve the 'real' interests of women? Was it an instance of empowerment, or did it actually disenfranchise the entire female population by depicting them as eternal victims? Foucault's answer would be to dismiss such questions as misplaced. Since women are not uniform beings, there is no way of knowing what the 'real' interests of women are, which makes the questions unanswerable. Instead, one should be investigating how the movement produced certain male and female identities. In this sense, power should not be understood as a repressive force that prevents women from pursuing their true objectives, but as a productive force afforded by digital platforms, contributing to define what counts as a 'true' man and a 'true' woman. Rather than understanding power and empowerment as synonyms for oppression and liberation, respectively, it is productive to think of them analytically as two sides of the same coin (Pease, 2002). The fact that power and empowerment can operate simultaneously does not mean that Foucault's conception of power is superior to the others. It merely means that the notion of subjectification allows us to identify power dynamics that may be obscured by the three other views. Similarly, understanding power in terms of coercion, manipulation, and domination allows us to understand practices that might otherwise be overlooked by the Foucaultian perspective – most notably instances of proper counter-power (see Ellerbrok, 2010) (Table 10.2).

Table 10.2 The four faces of power in organizations, as conceived by Fleming and Spicer (2014).

Power dimension	Definition	Blind spot
Coercion	The ability to force others to change their behavior	Instances where power is not immediately visible
Manipulation	The ability to prevent certain things from happening	Instances where power is embedded in societal institutions
Domination	The ability to influence the preferences of others	Instances of power that are more or less self-imposed
Subjectification	The process by which individuals are turned into subjects	Instances of counter-power

Power, empowerment, and sociomateriality

So, where does this lead us? Should we reject the concept of empowerment, often associated with digital organizing, completely? Or do we need to revise our conception of power and think of it as a multidimensional phenomenon that represents instances of oppression and empowerment as well as instances of subjectification? From a sociomaterial perspective, the latter would be the case. A sociomaterial perspective on power in digital organizing investigates different faces of power afforded by technology. In the same way that Zuboff (1988) and others have done, it is a matter of exploring how the material features of technology, as well as the cultural context in which these technologies are embedded, facilitates or enables a certain type of use, and how this usage manifests certain power relationships. Recently, a number of scholars have deployed the sociomaterial perspective to investigate how digital technology reconfigures power relationships in a range of organizations.

One example is Regula Burri's (2008) study of the introduction of new digital visualization technology to the medical field of radiology. Through ethnographic fieldwork, Burri observes how such technology has made it significantly easier for non-experts to read X-ray images, a development that challenges the expert status of radiologists. For instance, while the introduction of MRI scanners has empowered neurologists, cardiologists, and other occupational groups to use visualizing technology for their own purposes, it has also forced radiologists to rethink the boundaries of their own profession. While MRI has allowed some groups to get more 'things done' than previously, it has likewise altered the collective subjectivity of various professions. In a similar example, Ursula Plesner and Elena Raviola (2016) observe the introduction of new digital management devices in a public news corporation. One of these devices, an online platform displaying ongoing news stories to all members of the organization, has blurred the distinction between journalists and managers by significantly enhancing collaboration between the two groups. On the one hand, the introduction of shared digital platforms has empowered journalists to influence managerial decision-making about what stories to pursue and how. On the other hand, this same technology has challenged their professional independence by forcing them to continually negotiate journalistic priorities with managers and editors (for more examples, see Barrett et al., 2012; Petrakaki et al., 2016; Symon & Pritchard, 2015).

The overall point of these examples is that digital technologies cannot be seen exclusively as tools of empowerment (or 'technologies of freedom', as Castells calls them). They may be empowering in some instances, but less so in others. They may allow certain groups to challenge forces of coercion or manipulation, as in the case of the Arab Spring or the #MeToo movement, but this does not mean that these people escape the grip of power altogether. Power is omnipresent, as Foucault would say. The way that power relations are reconfigured in organizations is fully dependent on the sociomaterial context in which digital technology is submerged.

> **? QUESTIONS**
>
> 1. How do digital technologies lend themselves to traditional management control?
> 2. How do digital technologies enable empowerment? And what are the limits to such empowerment?
> 3. How do organizations use discourses of empowerment in their implementation of digital technologies?
> 4. How are personal and organizational identities shaped by the use of digital technologies?

References

Alvesson, M. & Willmott, H. (2002). Identity regulation as organizational control: Producing the appropriate individual. *Journal of Management Studies*, *39*(5), 619–644.

Andrews, F., Peigne, Y. & Vonberg, J. (2018). Catherine Deneuve denounces #MeToo in open letter. *CNN*, 11 January. Retrieved from http://edition.cnn.com/2018/01/10/europe/catherine-deneuve-france-letter-metoo-intl/index.html

Bachrach, P. & Baratz, M. S. (1962). Two faces of power. *American Political Science Review*, *56*(4), 947–952.

Bachrach, P. & Baratz, M. S. (1963). Decisions and nondecisions: An analytical framework. *American Political Science Review*, *57*(3), 632–642.

Ball, K. & Wilson, D. C. (2000). Power, control and computer-based performance monitoring: Repertoires, resistance and subjectivities. *Organization Studies*, *21*(3), 539–565.

Barnes, S.-A., Hoyos, M., Balduf, B., Behle, H. & Green, A. (2013). *Review of state of the art and mapping: Crowdemploy*. Warwick: Warwick Institute for Employment Research.

Barrett, M., Oborn, E., Orlikowski, W. J. & Yates, J. (2012). Reconfiguring boundary relations: Robotic innovations in pharmacy work. *Organization Science*, *23*(5), 1448–1466.

Bentham, J. (1791). *Panopticon or the inspection house*. London: T. Payne. Retrieved from https://archive.org/details/panopticonorins00bentgoog/page/n3

Benton, T. (1981). 'Objective' interests and the sociology of power. *Sociology*, *15*(2), 161–184.

Bergström, O. & Knights, D. (2006). Organizational discourse and subjectivity: Subjectification during processes of recruitment. *Human Relations*, *59*(3), 351–377.

Bernoff, J. & Schadler, T. (2010a). *Empowered: Unleash your employees, energize your customers, transform your business*. Boston, MA: Harvard Business Review Press.

Bernoff, J. & Schadler, T. (2010b). Empowered. *Harvard Business Review*, *88*(7/8), 94–101.

Braverman, H. (1974). *Labor and monopoly capital: The degradation of work in the twentieth century*. New York: Monthly Review Press.

Brett, D. (2015). *Digital nomad*. CreateSpace Independent Publishing Platform.

Brewis, J., Hampton, M. P. & Linstead, S. (1997). Unpacking Priscilla: Subjectivity and identity in the organization of gendered appearance. *Human Relations, 50*(10), 1275–1304.

Brooke, C. (2009). *Critical management perspectives on information systems*. Oxford: Butterworth-Heinemann.

Burawoy, M. (1979). *Manufacturing consent: Changes in the labor process under monopoly capitalism*. Chicago, IL: The University of Chicago Press.

Burri, R. V. (2008). Doing distinctions: Boundary work and symbolic capital in radiology. *Social Studies of Science, 38*(1), 35–62.

Castells, M. (2007). Communication, power and counter-power in the network society. *International Journal of Communication, 1*(1), 238–266.

Castells, M. (2009). *Communication power*. Oxford: Oxford University Press.

Castells, M. (2012). *Networks of outrage and hope: Social movements in the internet age*. Cambridge: Polity Press.

Cederström, C. & Spicer, A. (2017). *Desperately seeking self-improvement: A year inside the optimization movement*. New York: OR Books.

Chandler, D. (2017). All work and no pay: Creative industries freelancers are exploited. *The Guardian*, 18 May. Retrieved from https://www.theguardian.com/small-business-network/2017/may/18/all-work-and-no-pay-creative-industries-freelancers-are-exploited

Clegg, S. R. (1989). *Frameworks of power*. London: SAGE.

Clegg, S. R., Courpasson, D. & Phillips, N. (2006). *Power and organizations*. London: SAGE.

Coombs, R., Knights, D. & Willmott, H. C. (1992). Culture, control and competition: Towards a conceptual framework for the study of information technology in organizations. *Organization Studies, 13*(1), 051–072.

Costas, J. & Kärreman, D. (2013). Conscience as control: Managing employees through CSR. *Organization, 20*(3), 394–415.

Crozier, M. (1964). *The bureaucratic phenomenon*. Chicago, IL: Chicago University Press.

Dahl, R. A. (1957). The concept of power. *Behavioral Science, 2*(3), 201–215.

Dahl, R. A. (1961). *Who governs? Democracy and power in an American city*. New Haven, CT: Yale University Press.

Dickinson, E. (2011). The first WikiLeaks revolution? *Foreign Policy*, 3 January. Retrieved from http://foreignpolicy.com/2011/01/13/the-first-wikileaks-revolution/

DiMaggio, P. J. & Powell, W. W. (1983). The iron cage revisited: Institutional isomorphism and collective rationality in organizational fields. *American Sociological Review, 48*(2), 147–160.

Doolin, B. (1998). Information technology as disciplinary technology: Being critical in interpretive research on information systems. *Journal of Information Technology, 13*(4), 301–311.

Dyrberg, T. B. (1997). *The circular structure of power: Politics, identity, community*. London: Verso.

Edelman Intelligence (2017). *Crowdsourcing in America*. Retrieved from https://www.upwork.com/i/freelancing-in-america/2017/

Ellerbrok, A. (2010). Empowerment: Analysing technologies of multiple variable visibility. *Surveillance & Society, 8*(2), 200–220.

Fleming, P. & Spicer, A. (2014). Power in management and organization science. *The Academy of Management Annals*, *8*(1), 237–298.

Fortunati, L. (2014). Media between power and empowerment: Can we resolve this dilemma?. *The Information Society*, *30*(3), 169–183.

Foucault, M. (1977). *Discipline and punish: The birth of the prison*. New York: Pantheon.

Foucault, M. (1978). *The history of sexuality: The will to knowledge.* New York: Random House.

Foucault, M. (1980). *Power/knowledge: Selected interviews and other writings, 1972–1977.* New York: Pantheon.

Foucault, M. (1982). The subject and power. *Critical Inquiry*, *8*(4), 777–795.

Foucault, M. (2012). *The order of things: Archaeology of the human sciences.* London: Routledge. (Original work published 1966.)

Freire, P. (1970). *Pedagogy of the oppressed*. London: Continuum.

Gallie, W. B. (1955). Essentially contested concepts. *Proceedings of the Aristotelian Society*, *56*, 167–198.

Grint, K. & Woolgar, S. (1997). *The machine at work: Technology, work and organization.* Cambridge: Polity Press.

Hochschild, A. R. (2016). *Strangers in their own land: Anger and mourning on the American right.* New York: The New Press.

Honold, L. (1997). A review of the literature on employee empowerment. *Empowerment in Organizations*, *5*(4), 202–212.

Howe, J. (2008). *Crowdsourcing: How the power of the crowd is driving the future of business.* New York: Random House Business.

Husted, E. (2018). Mobilizing 'the alternativist': Exploring the management of subjectivity in a radical political party. *Ephemera: Theory and Politics in Organization*, *18*(4), 737–765.

Jones, R. (2000). Digital rule. *Punishment & Society*, *2*(1), 5–22.

Kanter, R. M. (1977). *Men and women of the corporation.* New York: Basic Books.

Kawasaki, G. & Fitzpatrick, P. (2014). *The art of social media: Power tips for power users.* New York: Portfolio.

Knights, D. & Morgan, G. (1991). Corporate strategy, organizations, and subjectivity: A critique. *Organization Studies*, *12*(2), 251–273.

Knights, D. & Willmott, H. (1988). *New technology and the labour process.* Houndsmill: Macmillan.

Kunda, G. (1992). *Engineering culture: Control and commitment in a high-tech corporation.* Philadelphia, PA: Temple University Press.

Laine, P.-M. & Vaara, E. (2007). Struggling over subjectivity: A discursive analysis of strategic development in an engineering group. *Human Relations*, *60*(1), 29–58.

Lukes, S. (1974). *Power: A radical view.* London: Red Globe Press.

Lukes, S. (2005). *Power: A radical view* (2nd edition). London: Red Globe Press.

Meyer, J. W. & Rowan, B. (1977). Institutionalized organizations: Formal structure as myth and ceremony. *American Journal of Sociology*, *83*(2), 340–363.

Michels, R. (1962). *Political Parties: A sociological study of the oligarchical tendencies of modern democracy.* New York: Crowell-Collier Publishing Company. (Original work published 1911.)

Pease, B. (2002). Rethinking empowerment: A postmodern reappraisal for emancipatory practice. *British Journal of Social Work*, *32*(2), 135–147.

Petrakaki, D., Klecun, E. & Cornford, T. (2016). Changes in healthcare professional work afforded by technology: The introduction of a national electronic patient record in an English hospital. *Organization, 23*(2), 206–226.

Plesner, U. & Raviola, E. (2016). Digital technologies and a changing profession: New management devices, practices and power relations in news work. *Journal of Organizational Change Management, 29*(7), 1044–1065.

Puffer, S. M. (2004). Introduction: Rosabeth Moss Kanter's 'Men and Women of the corporation and the change masters'. *The Academy of Management Executive, 18*(2), 92–95.

Reillier, L. C. & Reillier, B. (2017). *Platform strategy: How to unlock the power of communities and networks to grow your business*. London: Routledge.

Scarbrough, H. & Corbett, J. M. (1992). *Technology and organization: Power, meaning and design*. London: Routledge.

Sewell, G. & Wilkinson, B. (1992). 'Someone to watch over me': Surveillance, discipline and the just-in-time labour process. *Sociology, 26*(2), 271–289.

Shirky, C. (2008). *Here comes everybody: The power of organizing without organizations*. London: Penguin Books.

Socialmediatoday (2017). How I tripled my income in 3 months as a freelance digital marketer. Socialmediatoday, 14 February. Retrieved From https://www.socialmediatoday.com/marketing/how-i-tripled-my-income-3-months-freelance-digital-marketer

Sveningsson, S. & Alvesson, M. (2003). Managing managerial identities: Organizational fragmentation, discourse and identity struggle. *Human Relations, 56*(10), 1163–1193.

Symon, G. & Pritchard, K. (2015). Performing the responsive and committed employee through the sociomaterial mangle of connection. *Organization Studies, 36*(2), 241–263.

Townley, B. (1993). Foucault, power/knowledge, and its relevance for human resource management. *The Academy of Management Review, 18*(3), 518–545.

Upwork Blog (2017). A how-to guide for your perfect profile picture. Retrieved from https://www.upwork.com/blog/2014/10/how-to-guide-perfect-profile-picture/

Upwork Website (2017). Create an account. Retrieved from https://www.upwork.com/signup/

Zacharek, S., Dockterman, E. & Edwards, H. S. (2017). TIME person of the year 2017: The silence breakers. *TIME Magazine*, 18 December. Retrieved from http://time.com/time-person-of-the-year-2017-silence-breakers/

Zuboff, S. (1988). *In the age of the smart machine: The future of work and power*. New York: Basic Books.

11 IMPLICATIONS OF DIGITAL ORGANIZING

> With this book, we have suggested that it is fruitful to speak organizationally about digitalization (rather than digitally about organization). We have done so by engaging with concepts and concerns from a range of important organizational studies and theories. In this final chapter, we highlight some of the implications of digital organizing for individuals, organizations, and society that were raised in the previous chapters. Needless to say, this discussion is far from exhaustive. Also, when we talk about implications, we do not wish to speculate about the future, as such speculations often say more about the present than the future (Andersen & Pors, 2016). Rather, we will limit ourselves to the documented effects of digital organizing at the levels of individuals, organizations, and society. In that sense, this chapter is perhaps best viewed as a synthesis of the most important discussions raised in the second part of this book.

Implications for individuals

Digital organizing has different types of implications for different groups of individuals. It profoundly influences the tasks, relations, and identities of employees, and it alters the experience of being consumers or citizens who interact with organizations. In this section, we will consider some of these implications, beginning with issues of flexibility and the need to develop new approaches to work. We then discuss how individual privacy may be challenged by digital organizing in two senses: (1) when work extends beyond normal working hours, and (2) when the individual is made an object of tracking and surveillance. Finally, we touch on the experience of being a citizen in an increasingly digitalized public sector.

Deskilling or reskilling?

As shown multiple times throughout this book, the introduction of computers in the workplace during the 1980s had profound implications for the way people carried out their work. Shoshana Zuboff, a pioneer in this field of research, describes this change:

> The material alterations in means of production were manifested in transformations at intimate levels of experience – assumptions

about knowledge and power, [the workers'] beliefs about work and the meaning they derived from it, the content and rhythm of their social exchanges, and the ordinary mental and physical disciplines to which they accommodated in their daily lives. I saw that a world of sensibilities and expectations was being irretrievably displaced by a new world, one I did not yet understand. (Zuboff, 1988, p. xiii)

Some of the fundamental challenges to individuals in the workplace have to do with the demands for employee flexibility – the way in which employees are continually pushed to adjust to ever new work tasks and routines – and the implications for the kinds of skills needed to work with or alongside digital technologies. The constant theme in this discussion is whether digital organizing leads to a deskilling of employees because computers take over increasingly complex tasks, or whether it leads to a reskilling because employees learn to handle new kinds of work tasks, arising around the computerized parts of their work. In terms of the skills issue, Zuboff notes that "while it is true that computer-based automation continues to displace the human body and its know-how (a process that has come to be known as *deskilling*), the informating power of the technology simultaneously creates pressure for a profound *reskilling*" (Zuboff, 1988, p. 57).

Zuboff's conceptual distinction between 'the automating' and 'the informating' aspects of digital technologies, described in the previous chapter, has only become more relevant as organizations become increasingly digitalized: Although digital technologies perform many tasks – including highly sophisticated ones – they also produce a lot of data and new visibilities, to which employees and managers need to relate (Plesner & Justesen, 2018). Computers not only automate work, they also produce an extra layer of information about work processes. Whereas a common ambition is to make routine work easier and more efficient through digitalization, it is often overlooked that managerial work also becomes more complicated. For instance, access to and uses of data need to be closely managed, digital solutions need to be designed and maintained, and when digital technologies produce mistakes, someone has to take charge of fixing what needs to be fixed (Plesner & Justesen, 2018). As such, managing a digital organization is still hard work. For employees as well, detailed studies of work have shown how technologies that were supposed to make work easier are often "quite idiosyncratic, new failure modes appear continuously, and rote procedure cannot address unknown problems" (Orr, 2016, p. 2). This means that work-arounds, repair work, and improvisation around technology have become prevalent and time-consuming everyday activities for many employees.

As we have discussed several times throughout the book, the discourses of data-driven management, robotization, and artificial intelligence come with an implicit – or explicit – threat of eliminating jobs and handing over power to machines (Fleming, 2019). Digitalization has obviously made some jobs redundant and made digital technologies more powerful. However, the most important element to which we should pay attention at the level of the individual is probably how people's jobs are *changing*, and the kinds of new skills and routines that need to be learned and cultivated. By extension, it becomes relevant to inquire into people's subjective experiences of work and the emotions and values attached to working in a digital environment. It has been our argument throughout the book that the sociomaterial perspective provides a useful lens through which it is possible to consider both the 'objective' and the 'subjective' side of digital organizing.

The question of work–life balance

Another implication of digital organizing for individuals is related to what is commonly known as the 'work–life balance', an issue we touched on in Chapter 8 on communication and interactivity. The work–life balance discussion first arose in the mid-twentieth century, highlighting the need to formally separate work from leisure, and the importance of maintaining a healthy relationship between one's work life and one's private life. During the industrial era, work was typically both tedious and strenuous, which made the distinction between work and non-work – in factories represented by the punch clock – crucial to most people. However, with the rise of particularly cell phone technology, the distinction (and, hence, the balance) between work and private life gradually started to blur, leading to increased levels of stress and burnout among workers. As Kaspar Villadsen (2017, p. 365) notes:

> Employees are now working in cars, on weekends, in bed while watching TV, at their summer house watching the kids, sitting at Starbucks, waiting in line, on the train home. What we call 'work' may be carried out anywhere, any time, for a few minutes or for hours in a vacant time slot ... Workers may be constantly 'on call', available to superiors and colleagues, who in turn expect rapid responses and problem-solving, even during employees' time off. Turning off one's phone becomes a virtual act of negligence or insubordination.

That said, the dissolution of the work–life boundary may also have benefits for some individuals, most notably increased autonomy on the job or ability to work from home. In a study of so-called 'knowledge workers', Melissa Mazmanian and colleagues (2013) found that their interview respondents viewed the demand to be online 24 hours a day not as a matter of work encroaching on their personal life, but as a convenient opportunity to fulfill their own desire for professional autonomy and accountability. The employees' smartphones enabled them to check their emails during evenings or late at night, and this surely caused their physical and mental well-being to deteriorate. However, continual access also gave them the opportunity to work independently of space and time constraints as well as peace of mind in terms of being on top of things. There is thus a certain paradox to the autonomy afforded by digital organizing for these individual knowledge workers: on the one hand it allows them to work "anywhere/anytime", but it also forces them to work "everywhere/all the time" (Mazmanian et al., 2013, p. 1338).

Judy Wajcman similarly questions the common view that mobile devices lead to a colonization of employees' private lives. Through a large empirical study of how people actually use their phones, she found that they were primarily used for non-work-related issues throughout the day. According to an Australian nationwide survey, about 75 percent of all calls and 90 percent of all text messages are with employees' family and friends. Wajcman concluded that "by allowing some of the concerns of family and personal life to be handled during the working day, [mobile phones] might even be deployed to reduce time pressure" (Wajcman, 2015, p. 144). Empirical studies such as these allow us to question technological determinism in discussions of technology and work–life balance; mobile devices have various implications for the experience of work.

The surveilled and measured individual

No matter whether it is employees, managers, customers, or 'ordinary citizens', digital organizing makes measurement and surveillance an unavoidable part of their lives. Individuals are increasingly tracked on the basis of the digital traces they leave when they work, travel, shop, or search for information. We touched on this phenomenon in Chapter 7 on knowledge and datafication. Data about behavior and consumption can be used to tailor marketing and political campaigns to specific consumer groups and to monitor people's work productivity or shopping habits. Since organizations have realized the value of tracking and surveillance, it has been said – a bit polemically – that Google, for instance, knows you better than your best friends. The data sets that Google has about an individual are so massive that they can be used to predict the most specific consumption choices and shopping preferences. In Chapter 5 on production and produsage, drawing on Fuchs (2014), we referred to this as 'economic surveillance'.

In the context of work, surveillance has also become more pervasive. Returning to Zuboff, she observed already some 30 years ago that the informating dimension of digital technologies has produced a wealth of data on work processes that could be used to determine where mistakes were made and how individual employees performed. At the time, it seemed that through indirect ways of observing employees, that is through the analysis of data produced by digital technologies, managers could perform their work with less direct interaction and less personal investment (Zuboff, 1988). In other words, the machine took over the manager's watchful eyes.

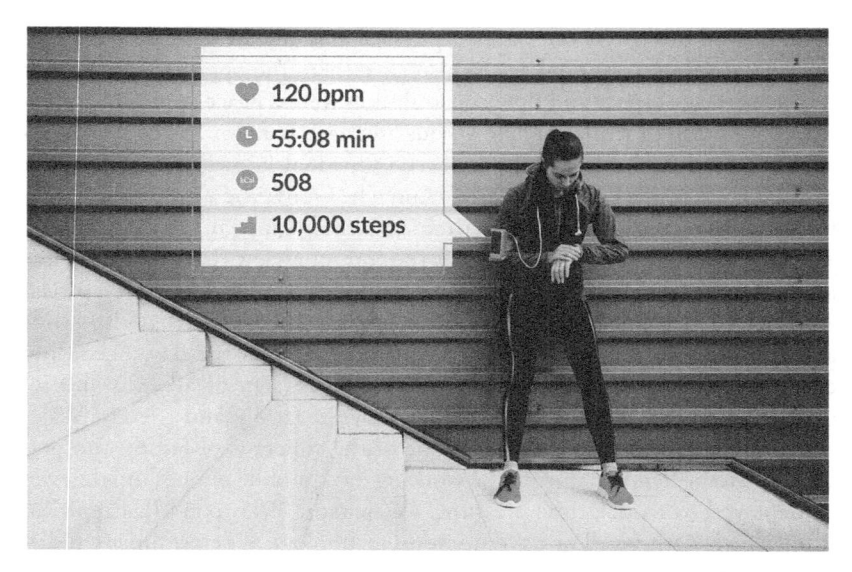

Image 11.1 Many people take advantage of the multiplicity of digital technologies that measure and track performance and health. This has given rise to the concept of 'the quantified self'.

Source: Todor Tsvetkov, iStock.

If we shift our attention to what it is like to be a consumer in relation to digital organizing, we can observe both an empowering trend and an exploitative one. As we noted in the discussions of co-creation and produsage, customers are often invited into development or production processes and acknowledged as capable of giving input and important to listen to. The interactive trend and the possibilities for engaging in debates about the choices of companies equally allow for customers' voices to be heard. At the same time, digital technologies also track consumer behavior and consumer preferences in order to profile them and predict their choices. In this way, consumer behavior and consumer profiles become data that can be exploited, traded, and repurposed. The business of trading data is immense, and although many people know that their every move is being registered, and that the ensuing data is sold, we see relatively little engagement from consumers with these aspects of digital organizing (Image 11.1).

Being a citizen in a digital public sector

Public sector digitalization is high on the agenda in many countries, and this means that when we as citizens interact with public sector organizations, we also experience the consequences of digitalization. Rather than interacting with public servants, we are asked to fill out applications online, and we may then expect our requests or cases to be processed by algorithms rather than people. In areas beyond ordinary administration, digital organizing also changes the experience of being at the receiving end of public services. For instance, public schools and public hospitals organize their teaching and their treatments around digital platforms, which *both* register pupils'/patients' data *and* give teachers/parents/patients/healthcare personnel access to data and thereby insight into the workings and knowledge of the public organization. In so far as citizens and public servants now relate to each other around digital platforms, we can see the emergence of a new kind of relationship between people and the bureaucracies that serve them. In a study of a digitalized public administration, this shift has been described as a move from face-to-face interaction to shoulder-by-shoulder interaction. The metaphor conveys the notion that citizens are being asked to help themselves online, in front of a screen, rather than standing in front of a civil servant with a desk between them (Pors, 2015).

The implications of digital organizing for individuals have been discussed in various strands of the organization literature, from those studying workplace surveillance and transparency (Ball, 2010; Orlikowski & Scott, 2014), to those concerned with privacy and the exploitation of our data (e.g. Heuer & Tranberg, 2013; Zuboff, 2019), to those preoccupied with the question of individuals' empowerment and choice (e.g. Ellerbrook, 2010). Individuals' work situations in relation to digitalization is also an issue for unions that are grappling with how to protect their members' interests, and for non-profit and non-governmental organizations (NGOs) such as Privacy International, which focuses on issues of data exploitation, surveillance, and privacy protection in relation to organizations and governments.

Implications for organizations

Throughout this book, we have given examples of how digital technologies become interwoven with organizational phenomena across different types of organizations. In this section, however, we will propose that some types of organizations are more profoundly impacted than others by certain dynamics. In this sense, we believe it makes sense to distinguish between private firms, public organizations, and voluntary organizations – which, while sharing many common concerns in relation to digitalization, might be affected differently by digitalization simply because they have different raisons d'être.

To begin with, it is worth noting that the discourse of disruption seems to affect all kinds of organizations – public, private, and voluntary – as many of them spend considerable resources ensuring that they are not left behind in what they see as an inevitable transformation of their domains due to digitalization. Organizations of whatever kind routinely hire consultants, join networks, attend future-oriented seminars, hire IT and social media specialists, invest in emerging technologies, and re-organize work around the latter. As Edward Schreckling and Christoph Steiger (2017, p. 15) put it in their rather deterministic exposé of these transformations: "Digitalization is no longer a choice but an imperative ... for all businesses across all industries and regions the motto is digitalize or drown". In this sense, it is probably fair to say that organizational design issues – which we discussed in Chapter 4 on structure and infrastructure – are central to all kinds of organizations. The issues raised are crucial: How should we redefine our work tasks? How should we rethink our work processes? How should we work with organizational development and strategy? Which types of employees do we need to hire?

Private firms

Besides such common concerns, it can be argued that for private firms, the so-called 'collaborative paradigm' (Gulbrandsen & Just, 2011) has pushed many firms to redefine how value is produced in digital environments, and to rethink work and labor processes as a field that extends beyond the boundaries of the organization proper. One example is the news industry, where the current focus on audience participation has disrupted, if not dissolved, the otherwise rigid boundary between journalists and their audience (Kammer, 2013). Beyond this industry, the possibility for corporations to exploit consumer engagement on digital platforms has allowed private firms to accomplish formerly expensive and time-consuming tasks rapidly and cheaply – the examples are numerous, as we showed in Chapter 6 on collaboration and co-creation.

For many private firms, an important aspect of digital organizing is to open up and search for value outside the organization, as expressed in the concepts of Open Innovation, Open Access, Open Strategy, and so on. At the same time, this opening up, and the ensuing blurring of boundaries between corporations and their environment, has led to complex strategy-making processes, in which digital technologies are an integral element (Plesner & Gulbrandsen, 2015). As strategy scholars Gianvito Lanzolla and Jamie Anderson (2008) argue, digitalization challenges organizations by enabling ubiquitous digital interaction with the organization's environment (the 'outside'). The authors describe how new media are pervasive, and how "companies have no choice but to leverage them" (Lanzolla & Anderson, 2008, p. 76). The perceived pressure from the outside compels organizations to strategically pursue new value chains and to try and control the digital platforms where interaction with customers takes place.

Power is another theme that must be considered in exploring how digital organizing affects private organizations. As we discussed in the previous sections, the new visibilities offered by digital technologies create new possibilities for control; sometimes these control possibilities end up reinforcing hierarchies, while under other conditions they erode them. Besides this effect of digital organizing on hierarchies, we saw in the previous chapter how identity and subjectivity are central to power. Studies of subjectivity within the digital economy have shown how multinational corporations play a leading role in the production of grand narratives about 'The Information Society'.

According to these studies, such narratives contribute to the production and mobilization of particularly active consumer-subjects (Moisander & Eriksson, 2006). From the perspective of a profit-making organization, we can observe how digital organizing contributes to the production of an ideal worker-subject who is flexible, adaptable, sociable, self-directing, and constantly able to commodify herself (Gill, 2010). Finally, and maybe most importantly, the aim of increasing profits has led many companies to pursue an aggressive automation agenda as a means of reducing labor costs (Roose, 2019). Also, many develop tracking and surveillance technologies that allow for the prediction of consumer behavior, thereby building valuable assets (Fuchs, 2010; Palmås, 2011; Zuboff, 2019). When rival firms use advances in digital technologies to pursue such strategies, many companies will feel compelled to do the same in order to stay competitive.

Public organizations

Public sector organizations are increasingly called on to develop more ambitious digitalization strategies. For example, public organizations are urged to explore possible uses of the Internet of Things in the design of smart cities, to use big data analyses to support decision-making, and to make different public digital platforms compatible. As we saw in the example of 'digitization-ready legislation' in Chapter 1, the ambition of making public organizations and public services increasingly digital leads to fundamental changes in the public sector. When laws need to be reviewed or revised in order to make sure that they are simple enough to allow for automation in case-processing, the result is a different type of public service and a different type of interaction between public authorities and citizens (Justesen & Plesner, 2018). Public organizations are perceived as legitimate only as long as they deliver a valuable service, and the role of digital technologies in this context is to provide more and better service for less taxpayer money. Public organizations face numerous challenges as they pursue this digital agenda. For instance, laws change regularly, and digital administrative systems need to be redesigned accordingly. Moreover, technologies evolve, and when a system changes in one part of the administration, it has implications for the interface with dependent systems in other parts of the administration. Finally, the complexity of the public sector and of citizens' needs challenges the digital requirements for standardization and automation. If complex cases are forced into standardized templates and handled by machines rather than public servants, citizens will often end up being treated poorly by the public authority, causing obvious legitimacy problems.[1]

As we discussed in Chapter 9, legitimacy issues are linked to transparency, and legitimacy is a particularly pressing issue for public organizations in the

---

1   This account is based on Plesner and Justesen's ongoing study of the digitalization of public organizations.

Image 11.2 The practice of hacking makes cyber security an increasingly important issue for public organizations. Accordingly, their work with transparency through open data needs to be handled with care.

Source: Gorodenkoff, iStock.

digital age. To appear legitimate, many organizations are called on to make their operations more transparent (Flyverbom, 2011). Government agencies publish previously hidden citizen databases online, public schools upload unprecedented amounts of information about children and teachers to designated learning platforms, and police officers are expected to carry small video cameras on helmets or in vehicles to show that they are not mistreating citizens. At the same time, however, many governments worldwide still conceal confidential information in the name of security, thus underscoring – once again – that transparency is never total and that it has to be manufactured and actively organized (Hansen & Flyverbom, 2015) (Image 11.2).

Public organizations also face specific challenges in utilizing the potentials of new knowledge production practices related to datafication. There are limits to which data – about citizens, for instance – public organizations are allowed to collect, store, use, and share across domains. Hence, public organizations are often less able than private firms to utilize the vast potential of big data as a source of knowledge creation. At the same time, they are increasingly expected to take advantage of such knowledge. The result is that they sometimes come to rely on privately owned platforms and technologies, posing a potential challenge to ideals of neutrality and democratic control (Madsen, 2015).

Voluntary organizations

In the case of voluntary (or non-profit) organizations, digital organizing plays a role not only for established NGOs, but also – very importantly – for more precarious organizations such as social movements and single-issue digital formations. All these types of organizations benefit from being present on digital platforms, in the sense that they acquire a potentially very wide reach (Castells, 2012). Moreover, online communities and special interest groups can both access and share information that was previously much more tightly controlled. The communicative dimension of digital organizing is thus an important benefit for most voluntary organizations. Numerous studies show how new information and communication technologies have significantly improved voluntary organizations' ability to mobilize support and advocate specific political causes. In fact, it has even been argued that the organizational structures of protest movements parallel the structures of digital platforms. As Michael Hardt and Antonio Negri remark in relation to revolutions that they call 'digital-platform driven':

> Much has been made of the way social media such as Facebook and Twitter have been employed in these encampments. Such network instruments do not create the movements, of course, but they are convenient tools, because they correspond in some sense to the horizontal network structure and democratic experiments of the movements themselves. Twitter, in other words, is useful not only for announcing an event but for polling the views of a large assembly on a specific decision in real time. (Hardt & Negri, 2011, n.p.)

This unique advantage, however, has also brought with it a range of challenges. Digital organizing can turn political action into a matter of personalized content sharing rather than collective action (Bennett & Segerberg, 2012) and thus detract from politics as an organizational phenomenon by allowing for the personalization of politics (Bennett, 2012). Studies focusing on organizational structures have likewise highlighted the potentially damaging effects of digital organizing. For instance, as Swann and Husted (2017) show, the Occupy Wall Street movement experienced a drift toward more hierarchical structures when the movement transformed from an offline protest camp and into an online movement centered on a designated Facebook page (see also Chapter 4, as well as Husted, 2015). Hence, to sustain the horizontal mode of coordination that typically characterizes social movements, it is necessary that such movements have a presence in both 'tweets and the streets' (Gerbaudo, 2012).

The issue of 'oligarchization' (i.e. the gradual centralization of power in the hands of a small elite) is not only relevant to social movements but also to other kinds of political organizations such as political parties. Robert Michels (1911) suggested that centralization of power constitutes an inescapable 'iron law' for all political parties, regardless of their otherwise democratic aspirations. However, with the emergence of digital technology, a new wave of parties such as *Movimento 5 Stelle* in Italy, *Podemos* in Spain, and the *Pirate Party* in Northern Europe have started to employ digital technology to actively counter the process of oligarchization. Gerbaudo (2018) refers to these organizations as 'digital parties', in order to highlight their reliance on online platforms for mobilizing support, coordinating events, and facilitating decision-making processes.

Despite good intentions, however, many of these parties eventually end up reproducing some of the century-old dynamics of traditional political parties. For instance, in a study of the Danish green party *Alternativet*, Emil Husted and Ursula Plesner (2017) show how a designated platform for online deliberation affords disengagement from party members by encouraging them to voice their opinion on a number of issues but without granting them the mandate to edit or discard the content of specific policy proposals. Similarly, Gerbaudo (2018, p. 76) uses the term 'distributed centralization' to describe situations where digital parties encourage their members to take the initiative in carrying out various projects, but where the overall strategy is predetermined by the party leadership. With digital parties, we thus get a type of oligarchization that is cloaked in a veil of participation and engagement, but where the party leadership ultimately retains much the same power as in traditional parties.

Finally, identity is an organizational issue, which is important to consider when exploring practices of digital organizing in voluntary organizations. For instance, so-called 'hybrid mobilization movements' (Chadwick, 2007) such as MoveOn and Avaaz have demonstrated a remarkable ability to strategically oscillate between different organizational identities. By successfully merging the identities of a political party, an interest group, and a social movement, these organizations are able to attract an incredibly wide range of supporters. While this kind of convergence has contributed to the success of such organizations in assembling networks of millions of members worldwide, it has likewise made it difficult for the organizations to articulate a coherent voice. This, in turn, might be compensated by these organizations' strategic use of different digital platforms such as YouTube, Facebook, or MySpace (Kavada, 2012).

Implications for society

In the book's first chapter, we described computational power as growing exponentially, and pointed out how, together with algorithmic thinking and connectivity, the growth of computational power has been the foundation of a thoroughgoing digitalization of organizations. The developments we have described here have not only had implications for individuals and organizations, but also at the societal level. Various national and global institutions are grappling with regulatory issues and a number of side effects related to digital organizing. At the core of the efforts to develop new national policies and transnational agreements regarding digitization lies an emerging recognition that the large corporations, which deliver important parts of the digital infrastructure – and capitalize heavily on it – are altering the world's balance of power.

Every year, *Forbes Magazine* publishes a list of the world's most powerful men and women, and every year, the list is headed by the president of a large country. Four years in a row, Vladimir Putin came out on top, curtailed by the leaders of the USA, China, and Germany. To most people, this makes sense. World leaders make important decisions that influence many people, which is what makes them powerful. However, a few steps down the list, we discover more unusual characters. For instance, in 2016, Larry Page (co-founder of Google) and Mark Zuckerberg (co-founder of Facebook) made it to the top ten. Clearly, the founder of a tech-conglomerate has the ability to influence a whole lot of people, but why was Larry Page considered more powerful than Narendra Modi, the prime minister of India? And how come Mark Zuckerberg was considered more powerful than Theresa May, the former prime minister of Great Britain? To clear things up, *Forbes* published a short statement that outlines the selection criteria behind the list. It reads as follows:

> To compile the list, we considered hundreds of candidates from various walks of life all around the globe, and measured their power along four dimensions. First, we asked whether the candidate has power over lots of people. Pope Francis, ranked #5, is the spiritual leader of more than a billion Catholics ... Next we assessed the financial resources controlled by each person. Are they relatively large compared to their peers? ... Then we determined if the candidate is powerful in multiple spheres ... Lastly, we made sure that the candidates actively used their power. North Korean dictator Kim Jong-un (#43) has near absolute control over the lives of the 25 million people who live in his country, and is known to punish dissent with death. (Ewalt, 2016, n.p.)

This statement reveals at least three things about *Forbes*' list of powerful people, which we can connect back to our discussions of power and empowerment in Chapter 10. First, it illustrates that power is conceived as something that can be possessed. Pope Francis *has* a lot of power, because he is the spiritual leader of more than a billion people. Second, the statement shows that power is viewed as something that is done to someone. In the case of Kim Jong-un, power is the ability to exercise absolute control *over* millions of North Korean citizens. Finally, the statement suggests that the pioneers of digital technology are powerful both because of their financial resources and perhaps because of their ability to *empower* lots of people. The products they offer have proven so useful and so popular that their companies have experienced explosive growth. On the basis of their size, they are able to pursue aggressive market leader and market dominance positions, just like they are able to generate unfathomable wealth.

Because the tech companies are global corporations, and because regulatory bodies have a hard time keeping up with their activities, they have managed to only pay negligible taxes, to refuse taking responsibility for the uses of their infrastructures, to exploit cheap labor, and to play out nations against one another in their competition to make it attractive for the big tech companies to establish themselves in a given country. We list these issues in order to indicate that many of the affordances of digital infrastructures and much of the infatuation with digital technologies and digitalization have overshadowed the concerns that nations and the global community can rightfully raise in the face of digitization. These concerns range over a broad spectrum: power, market dominance, economy, exploitation, and ethics.

Among the many critiques and public debates regarding the implications of digital organizing on a societal level, one of the most fundamental analyses of a new societal époque has been delivered by Zuboff (2019) in her book *The Age of Surveillance Capitalism: The Fight for a Human Future at the New Frontier of Power*. As the title indicates, Zuboff's book is not solely about the promises and pitfalls of digital technologies, it is a description of a new phase in the evolution of capitalism. She dates the beginning of this era to 2001, and argues that Google was a driving force in initiating it. In the early days, Google searches were not used for anything. Zuboff describes how, at the time, they called the information they had about users' search patterns 'data exhaust', indicating that their search track was simply a by-product of their real service, namely information search. At a certain point, however, Google chose to give up their initial resistance against advertisements and use their exclusive access to user data logs to generate income from ads. Google had realized that their 'data exhaust' – or the by-product of internet searches – was extremely valuable, and other tech companies began to follow suit, voraciously collecting user data and selling it.

As described in Chapter 7 on knowledge and datafication, the value of this search data comes from the possibility of analyzing the behavioral traces that people leave when they interact on digital platforms and then use these traces to predict future behavior. Such predictions can be used to tailor messages and target individuals very effectively – as consumers, voters, or in other capacities. Today, it is not just tech companies that use online surveillance to generate value – it has become a standard model, also in sectors such as insurance, healthcare, transportation, education, and entertainment. In Zuboff's view, surveillance capitalism is based on the exploitation of *human nature*, just like industrial capitalism depended on the exploitation of *nature*. People willingly give away their behavioral data, often because they are offered 'smart', personalized service, thereby contributing to a new economy and new power relations, where those who have the most information 'have' the power. Relatedly, Martin More (2019) has described how the harvesting of data and the personalization of marketing allows for very precise – and manipulative – messages that undermine democratic debates: nihilist subcultures generate false stories just for fun, economic elites invest in the manipulation of voters through micromarketing, and foreign powers destabilize national elections and democratic debates through hacking, bots that spread false stories, and so on.

Other grand challenges have to do with new north/south relations, as digital platforms allow for a new wave of exploitation of cheap labor – for instance, when people in affluent areas of the world save money by offering ad hoc tasks like language editing online, to get services done cheaply without having to worry about social security. Another example of the easy exploitation of cheap labor online can be seen in a gaming phenomenon like gold farming. Gold farming refers to the practice of laborers in developing countries who play multiplayer online games in order to acquire in-game currency and sell it for real-world money to people who have less time to play themselves. Finally, if we extend our thinking about global inequalities related to digital organizing, an obvious problem is how some nations succeed in censoring the internet, and some nations fail to offer the same digital infrastructures to all their citizens. In this way, inequalities are amplified through differences in access (see also Chapter 5).

In the face of these challenges, the question of regulation arises. It is hotly debated how much should be regulated and how much should be left to the market. After all, a large part of the developments described above are based on people's voluntary contributions to an economy that rests on digitalization. One camp affirms the value of free speech and the freedom of anyone to capitalize on our basic, shared digital infrastructure, namely the internet. But there is another camp that insists on trying to limit the economic and political ramifications of the dominance of the tech giants. For instance, the European Commission fined Google €2.4 billion for monopolistic behavior, platforms infringing on intellectual property rights have been shut down, the operating power of sharing platforms has been curtailed by local legislation, and so on. Perhaps it is not unfair to compare the societal challenge of digitalization to that of climate change, as did *The Guardian* commentator Charles Arthur (2017, n.p.):

> Even as they revel in their network-reinforced positions, the big tech companies are battling with problems so big and intractable, and so far-reaching in their effects, that to find comparisons in the real world you have to look for truly global phenomena. The problems engendered by the internet have crept up on us over the years, but only recently have they seemed overwhelming. It's like a social form of climate change, and the analogy works surprisingly well.

Throughout this book, we have strived to show how digital technologies are interwoven with various aspects of organizing; we have argued that digital technologies do not lead to a complete transformation of all aspects of organizations. It is not as if "all our old assumptions" no longer "fit the facts" (Toffler, 1980, p. 16), and that we therefore need to free ourselves of "the baggage of 'common sense' assumptions and understandings" (Bruns, 2008, p. 2) about organizational dynamics, as some techno-optimists argue. Rather, classical organizational themes and concepts remain relevant in analyzing how organizations digitalize various aspects of their operation. We have discussed the classical themes and concepts in relation to more contemporary ones in order to show how they complement each other. Our argument is that bringing together the classical and the new themes allows us to think organizationally about digitalization.

With this chapter, we close the book with pointing to some rather dramatic implications of digital organizing for individuals, organizations, and society. We do not wish to participate in the otherwise omnipresent disruption discourse. Rather, we wish to encourage reflections about how we would like to be treated as individuals – and treat other individuals – when we organize around digital technologies. We wish to inspire ethical discussions about the uses of data and automation in organizations. And we wish to stimulate discussions about what kind of society we want to live in – a society which will certainly be enveloped by digital technologies. Digital technologies are interwoven with organizations, but not in any necessary ways. In the discussions throughout this chapter, we have identified a number of dilemmas. We need to have a language for discussing these dilemmas, and to thoroughly consider their implications for individuals, organizations, and society (Table 11.1).

Table 11.1 Dilemmas of digital organizing.

Individuals	Deskilling versus reskilling Smart work versus compensation work Autonomous work versus 'always on' Being seen versus being surveilled
Organizations	Value of openness versus cost of openness Expectations to digitalize versus cost of digitalizing Easier mobilization versus risk of fragmentation
Society	New value-adding infrastructures versus power concentration New opportunities for division of labor across the globe versus reinforcement of inequality

References

Andersen, N. Å. & Pors, J. G. (2016). *Public management in transition: The orchestration of potentiality*. Bristol: Policy Press.

Arthur, C. (2017). Internet regulation: Is it time to rein in the tech giants? *The Guardian*, July 2. Retrieved from https://www.theguardian.com/technology/2017/jul/02/is-it-time-to-rein-in-the-power-of-the-internet-regulation

Ball, K. (2010). Workplace surveillance: An overview. *Labor History*, 51(1), 87–106.

Bennett, W. L. (2012). The personalization of politics: Political identity, social media, and changing patterns of participation. *The ANNALS of the American Academy of Political and Social Science*, 644(1), 20–39.

Bennett, W. L. & Segerberg, A. (2012). The logic of connective action: Digital media and the personalization of contentious politics. *Information, Communication & Society*, 15(5), 739–768.

Bruns, A. (2008). *Blogs, Wikipedia, second life, and beyond*. New York: Peter Lang.

Castells, M. (2012). *Networks of outrage and hope: Social movements in the internet age*. London: Polity Press.

Chadwick, A. (2007). Digital network repertoires and organizational hybridity. *Political Communication*, 24(3), 283–301.

Ellerbrook, A. (2010). Empowerment: Analysing technologies of multiple variable visibility. *Surveillance & Society*, 8(2), 1477–7487.

Ewalt, D. M. (2016). The world's most powerful people 2016. *Forbes Magazine*, December 14. Retrieved from https://www.forbes.com/sites/davidewalt/2016/12/14/the-worlds-most-powerful-people-2016/#7b5340451b4c

Fleming, P. (2019). Robots and organization studies: Why robots might not want to steal your job. *Organization Studies*, 40(1), 23–38.

Flyverbom, M. (2011). *The power of networks: Organizing the global politics of the internet*. Cheltenham: Edward Elgar.

Fuchs, C. (2010). Web 2.0, prosumption, and surveillance. *Surveillance & Society*, 8(3), 288–309.

Fuchs, C. (2014). *Digital labour and Karl Marx*. London: Routledge.

Gerbaudo, P. (2012). *Tweets and the streets: Social media and contemporary activism*. London: Pluto Press.

Gerbaudo, P. (2018). *The digital party: Political organisation and online democracy*. London: Pluto Press.

Gill, R. (2010). Life is a pich: Managing the self in new media work. In M. Deuze (Ed.), *Managing media work* (pp. 249–262). London: SAGE.

Gulbrandsen, I. T. & Just, S. N. (2011). The collaborative paradigm: Towards an invitational and participatory concept of online communication. *Media, Culture & Society*, 33(7), 1095–1108.

Hansen, H. K. & Flyverbom, M. (2015). The politics of transparency and the calibration of knowledge in the digital age. *Organization*, 22(6), 872–889.

Hardt, M. & Negri, A. (2011). The fight for "real democracy" is at the heart of Occupy Wall Street. *Foreign Affairs*, October 11. Retrieved from https://www.foreignaffairs.com/articles/north-america/2011-10-11/fight-real-democracy-heart-occupy-wall-street

Heuer, S. & Tranberg, P. (2013). *Fake it! Your guide to digital self-defense*. Scotts Valley, CA: Amazon Books.

Husted, E. (2015). From creation to amplification: Occupy Wall Street's transition into an online populist movement. In J. Uldam & A. Vestergaard (Eds.), *Civic engagement and social media: Political participation beyond protest* (pp. 153–173). Basingstoke: Palgrave Macmillan.

Husted, E. & Plesner, U. (2017). Spaces of open-source politics: Physical and digital conditions for political organization. *Organization*, 24(5), 648–670.

Justesen, L. & Plesner, U. (2018). Fra skøn til algoritme: Digitaliseringsklar lovgivning og automatisering af administrativ sagsbehandling. *Tidsskrift for Arbejdsliv*, 20(3), 9–23.

Kammer, A. (2013). Audience participation in the production of online news: Towards a typology. *Nordicom Review*, 34, 113–126.

Kavada, A. (2012). Engagement, bonding, and identity across multiple platforms: Avaaz on Facebook, YouTube, and MySpace. *MedieKultur: Journal of Media and Communication Research*, 28(52), 28–48.

Lanzolla, G. & Anderson, J. (2008). Digital transformation. *Business Strategy Review*, 19(2), 72–76.

Madsen, A. K. (2015). Between technical features and analytic capabilities: Charting a relational affordance space for digital social analytics. *Big Data & Society*, 2(1), 1–15.

Mazmanian, M., Orlikowski, W. J. & Yates, J. (2013). The autonomy Paradox: The implications of mobile email devices for knowledge professionals. *Organization Science*, 24(5), 1337–1357.

Michels, R. (1962). *Political parties: A sociological study of the oligarchical tendencies of modern democracy*. New York: Crowell-Collier Publishing Company. (Original work published 1911.)

Moisander, J. & Eriksson, P. (2006). Corporate narratives of information society: Making up the mobile consumer subject. *Consumption Markets & Culture*, 9(4), 257–275.

Moore, M. (2019). *Democracy hacked: Political turmoil and information warfare in the digital age*. London: Oneworld Publications.

Orlikowski, W. J. & Scott, S. V. (2014). What happens when evaluation goes online? Exploring apparatuses of valuation in the travel sector'. *Organization Science*, 25(3), 868–891.

Orr, J. E. (2016). *Talking about machines: An ethnography of a modern job*. Ithaca, NY: Cornell University Press.

Palmås, K. (2011). Predicting what you'll do tomorrow: Panspectric surveillance and the contemporary corporation. *Surveillance & Society*, 8(3), 338–354.

Plesner, U. & Gulbrandsen, I. T. (2015). Strategy and new media: A research agenda. *Strategic Organization*, 13(2), 153–162.

Plesner, U. & Justesen, L. (2018). Merproduktion af målbarhed: Synlighed og nye ledelsesopgaver. *Tidsskrift for Arbejdsliv*, 20(4), 108–115.

Pors, A. S. (2015). Becoming digital: Passages to service in the digitized bureaucracy. *Journal of Organizational Ethnography*, 4(2), 177–192.

Roose, K. (2019). The hidden automation agenda of the Davos elite. *The New York Times*, January 25. Retrieved from https://www.nytimes.com/2019/01/25/technology/automation-davos-world-economic-forum.html

Schreckling, E. & Steiger, C. (2017). Digitalize or drown. In G. Oswald & M. Kleinemeier (Eds.), *Shaping the digital enterprise* (pp. 3–27). Cham: Springer.

Swann, T. & Husted, E. (2017). Undermining anarchy: Facebook's influence on anarchist principles of organization in Occupy Wall Street. *Information Society*, 33(4), 192–204.

Toffler, A. (1980). *The third wave.* London: Pan Books.

Villadsen, K. (2017). Constantly online and the fantasy of "work–life balance": Reinterpreting work-connectivity as cynical practice and fetishism. *Culture and Organization, 23*(5), 363–378.

Wajcman, J. (2015). *Pressed for time: The acceleration of life in digital capitalism.* Chicago and London: The University of Chicago Press.

Zuboff, S. (1988). *In the age of the smart machine: The future of work and power.* New York: Basic Books.

Zuboff, S. (2019). *The age of surveillance capitalism: The fight for a human future at the new frontier of power.* New York: PublicAffairs.

INDEX

A

Abstract events, 46
Actor-network theory (ANT), 49, 66–67
Adhocracy, 87, 88
Advanced Research Projects Agency Network (ARPANET), 11–12
Affordances, 68–73
   analysis implications, 72–73
   association, 70
   concept of, 68–69
   digital technologies, 187
   editability, 70
   persistence, 70
   PowerPoint technology and, 69
   social media and, 69–70, 192
   visualizations and, 70
Agency for Culture, Denmark, 149
Age of Surveillance Capitalism: The fight for a human future at the new frontier of power (Zuboff), 252
Age of the Smart Machine, In the (Zuboff), 97, 225
Agile organizing and sprints, 40
Albu, O. B., 175–176
Algorithmic big data analysis, 209
Algorithmic thinking
   artificial intelligence, machine learning and, 10
   binary logics, digitization-ready legislation and, 5–7
   challenges/weaknesses of, 7–10
   connectivity and, 11–12
   decision-making, delegating to machines, 10–11
   history/definition of, 4–5
   transistor development and, 9–10
Algorithms, big data and, 169
Alienation
   technology-dependent, 57
   theory, Marx, 112
Amazon, 124, 165–166, 169
America on Line (AOL), 24, 106, 107
Anderson, Chris, 60, 120, 166

Anti-essentialism, 62–63
Apple Inc., 16, 110, 144–146
Arab Spring, 233–234, 235
Argyres, Nicholas, 95
ARPANET (Advanced Research Projects Agency Network), 11–12
Artifacts, organizations and, 184
Artificial intelligence (AI), 54
   big data and, 167
   blended automation, 19–21
   bounded automation and, 17–19
   machine learning and, 10
   Watson, IBM project, 158
Assembly lines, 57, 137, 225–226
Association affordance, 70
Automating, informating *v.*, 241
Automation
   bounded, 17–19
   power, smart machines and, 225–226
   smart machines and, 225–226
Avaaz, 250

B

Babbage, Charles, 4
Bachrach, P., 221–222
Bamforth, Ken, 38–39
Bannon, L. J., 139–140
Baratz, M. S., 221–222
Behavioral economists, 134
Bell, Alexander Graham, 63
Bentham, Jeremy, 226–227
Berners-Lee, Tim, 12
Bernoff, J., 231, 232
Bernstein, E. S., 205
Best Buy, 231, 232, 233
Best fit structure, 88
Bethlehem Steel Company, 31
Beunza, Daniel, 19
Bias, ideological, 197
Big data
   analysis, predictions/monitoring displays of, 165–166, 209
   generation/uses of, 164–165

INDEX

knowledge and, 166–168
non-transparency in police work case, 169–170
real-time data to actionable knowledge, 168–169
transparency and, 210
Binary logics, digitization-ready legislation and, 5–7
Blau, Peter, 82, 87
Blauner, Robert, 57, 58
Blended automation, 19–21
Blockbuster, 84
Blockchain
    connectivity and, 13–14
    virtual organization and, 96
Blockchain for Dummies (Gupta), 13
Blogging, 106–107, 197
Blogs, Wikipedia, Second Life, and Beyond (Bruns), 118
Boolean algebra, 4
Boole George, 4
Boundaries, social media and organizational, 189
Bounded automation, 17–19
Bowker, Geoffrey C., 83
Braverman, Henry, 58, 109n1, 111–113, 116, 126–127
Bridgeman, Todd, 64
British Library, 88
British National Health Service (NHS), 64
Bruns, Axel, 60, 118–120
Brynjolffsson, Erik, 10–11
Burawoy, Michael, 113
Bureaucracy, organizational structure, 86–87
Bureaucratic enterprise, traditional design, 85
Bureaucratic organization approach, 33–34
Burri, R. V., 235
Business communication, 180
Business Insider, 107
Business Process Re-Engineering (BPR), 85

C

Cambridge Analytica, 73
Capital accumulation, 111
Capitalism, 252
Capital (Marx), 55
Carey, J. W., 139
Castelló, I., 204

Castells, Manuel, 90, 115, 232–233
Champy, James, 85
Chandler, Alfred D., 84–85
Chaplin, Charlie, 32
Chia, Robert, 47
Chinese factory workers, 126
Christensen, Clayton, 16
Citizen, in digitized public sector, 244–245
Classical model of corporate legitimacy, 198, 199
Classical organizational theory, 33, 81–82
Classroom, as learning platform, 206
Clegg, Stewart, 43, 135, 220, 222
Co-construction, 132
Co-creation, 22, 24
    collaboration and, 131–133
    defining, 132
    digital platforms, Nike and, 144–146
    openness paradigm and, 142–144
    organizing crowds digitally, Innosite, 146–148
    public sector and, 146
    sociomateriality, collaboration and, 148–150
Coercion
    power in organizational studies, 220–221, 234
    subjectification and, 231–232
Cole, Robert E., 156, 157, 161
Collaboration
    behavior and, 134
    co-creation and, 132–133
    competition and, 135
    as design issue, 135–136
    digital technologies, software for, 139
    economic origins, 134–135
    innovation and, 138
    lifecycle, 136
    online, 94–96
    organizations, between, 131–132
    sociomateriality, co-creation and, 148–150
    web-based technologies for, 140–141
Collaborative arrangements, phases in, 136–137
Communal evaluation, produsage and, 119–120
Communication
    corporate, 180–181
    institutionalization of, 176–178

259

INDEX

Communication (*Continued*)
 interactivity, and, 184–191. *See also* Interactivity, communication and
 internal/external, 178
 material approaches to, 183–184
 narrative approaches to, 182–183
 new approaches to, 181–182
 strategic, 179–180
Communication, interactivity and
 assumptions of, 185
 challenges to controlled communication, 175–176
 defining/importance of, 174–175
 social media and, 186
 sociomateriality and, 191–192
 See also Interactivity, communication and
Communication as Constitutive of Organization (CCO), 183
Communication process, 178
Communication studies, 176–178
Communication theory, interactivity and, 185–186
Communities of practice, 137
Competition, collaboration and, 135
Complexity, new technologies, 45
Computational thinking, 7
Computer-aided manufacturing systems, 114
Computer-based technology, automation and, 226
Computer punch card, 6
Computer-Supported Cooperative Work (CSCW), 139–140
Connectivity
 blockchain and, 13, 14
 digitalization of interactions and, 11–12
 ethos of openness, Internet and, 12–14
 interconnectedness of computers and, 11–12
 material issue, 14–15
Consumer
 feedback loop, producers and, 118
 industrial age and, 117. *See also* Prosumers, production and
 Internet and, 117–118
 passive, 122
Contingency, organizational studies and, 39–43
 defined, 40

fitness hypothesis and, 40
structure and, 84
task-related technology and, 41–42
theory of, 40
variability/analyzability and, 42–43
Continuous events, 45, 46
Conversational Firm, The (Turco), 89
Coombs, Rod, 64
Cooper, Geoff, 61–62
Cooperative work, computer supported, 139–140.
Co-production, 175–176, 185
Core concepts, organizational communication perspectives, 191
Corporate blogging, 197
Corporate communication, 180–181
Corporate legitimacy
 models of, 198–199
 See also Legitimacy
Corporate messages, interactive communication and, 175–176
Corporate reputation, transparency and, 205
Corporate Social Responsibility (CSR), 201
Counter-power
 defining, empowerment *v.*, 232–233
 in practice, #MeToo case, 233–234
Creative Commons license, 120
Crowdsourcing
 Innosite case, 146–148
 rankings/reviews, 212
Crowdsourcing: How the power of the crowd is driving the future of business (Howe), 230
Crozier, M., 221
Culture, organizational, 225
Cunliffe, Ann, 40
Customer engagement, Nike, 145–146
Cyber proletariat, produsage and, 125–126
Cyber-Proletariat (Dyer-Witheford), 125

D

Dahl, Robert, 220–222
Data analytics, organizational decision-making and, 162–164
Database, common access, 95
Data centers, 14–15
Data exploitation, 245

INDEX

Datafication, 22, 25
   data-driven decision-making, 163–164, 170
   knowledge, home care practice and, 154–156
   new organizational practices, 163
   as organizational driver, 162–163
   sociomateriality, knowledge and, 170–171
Decision-making
   algorithmic, 6
   data-driven, 162–164, 170
   machine, delegating to, 10–11
Democratization, organizational, 39
Denyer, D., 140–141
Design
   collaboration as organizational issue, 135–136
   organizational, 84–85
Deskilling, reskilling v., digital organizing, 240–242
Determinism, technological, 54–55
Difference Engine, 4
Digital communications technologies, communication, interactivity and. *See* Communication, interactivity and
Digital economy, beneficiaries of, 124–125
Digital empowerment, 229–232, 233–234
Digital infrastructures
   bureaucracy and, 86
   defined, 83
   social media, organization and, 101–102
Digitalization
   blended automation, 19–21
   bounded automation, 17–19
   of classrooms, 206
   defined, 7
   development of, 15–16
   digital technologies, jobs and, 242
   entertainment industry, 84
   health professionals and, 100
   implications, organizations, 245–246
   interactions, computer, 11–12
   organizational process, 19
   power and, 246
   private firms and, 246–247
   public organizations and, 247–248
   public sector and, 244–245
   transparency, legitimation and, 206
   voluntary organizations and, 248–250
Digital learning platform, transparency and, 206–208
Digital organizing
   citizens, public sector, 244–245
   crowdsourcing, Innosite, 146–148
   deskilling/reskilling, 240–242
   dilemmas of, 254
   literature, 230
   political action and, 249
   social uprisings and, 233–234
   society, implications, 250–254
   sociomateriality and, 235
   surveilled individual, measured v., 243–244
   virtual library, 82–83
   work-life balance, 242–243
Digital platforms
   co-creation, Nike and, 144–146
   collaborations and, 139
   dimensions/extremes, collaboration and, 147
   interactive communications and, 191–192
   Internet and, 117–118
   journalists and, 235
   manager and, 148
   organizing crowds, Innosite, 146–148
   social media, communication interactivity and, 187–188
   stakeholder communications and, 203–204
   surveillance technology, 206–207
   UpWork freelance work, 218–220
   web-based technologies, 140–141
Digital technology
   affordances, 187
   collaborative software, 139
   discipline, supervise at distance, 228
   environmental cost of, materials, 125–126
   interactivity and, 190
   knowledge datafication, home care practice and, 154–156
   knowledge management systems (KMS) and, 161
   managerial control/transparency and, 210–212
   networked organization and, 92
   performance tracking, 244
   power/empowerment and, 235

INDEX

Digital technology (*Continued*)
  terminology of, 21–22
  workplace, functions of in, 229
  See also Technology
Digitization, 7
Digitization-ready legislation, binary logics, 5–7
Dimaggio, P. J., 223
Discipline and Punish (Foucault), 227
Discourse, communication *v.*, 182
Displays, monitoring big data, 165–166
Disruptive innovation, 16
Divisionalized form, 87–88
Division of labor, 137
Domain Awareness System, 169
Domination, power and, 222–223, 234
Driverless car, 10–11
Drucker, Peter, 157
Duguid, P., 138
Dyer-Witheford, Nick, 125–126

E

Economic knowledge, 135
Economics
  collaboration and, 134–135
  knowledge and, 156–157
Economic surveillance, 124
Economy, collaborative infrastructure and, 117
Editability affordance, 70
Efficiency
  communications and, 177
  increasing, industrial manufacturing, 109
  time-and-motion studies, 31–32
Empowered: Unleash your employees, energize your customers, and transform business (Bernoff & Schadler), 231
Empowerment
  counter-power, 232–233
  digital, 229–232
  employees and, 217–218, 231–232
  #MeToo, 233–234
  sociomateriality, power and, 235
  UpWork freelance platform, 218–220
  See also Power
Encrypted transaction technology, 13, *See also* Blockchain *entries*
Enforced standardization, 31–32
Engineering, scientific management and, 31
Engineering technologies, 42

Enlightenment, 43, 47, 59, 227
Enterprise 2.0, 140–141
Enterprise Social Media, 99, 189
Entertainment industry, digitalization and, 84
Environment
  digital technology, materials and, 125–126
  knowledge intensive, collaboration and, 138
Equivocality
  defined, 44
  new technologies and, 44–45
  organizational studies and, 44–46
Espeland, W. N., 208–209
Ethos of openness, connectivity, 12–14
Etter, M., 175–176
Evaluative process, legitimacy as, 199–201
Experience-based knowledge, 157
Explicit knowledge, 160
Exploitation, workers, 124, 252–253
Exponentiality, digitalization and, 9–10
External communication, 178

F

Facebook, 72, 101, 117, 123, 124, 146, 187–189, 209, 249
Facebook Revolutions, 233n4
Face-to-face interactions, 141, 148–149
Fayol, Henri, 31
Feedback loop, consumer/producers, 118–119, 211, 231
Fitness hypothesis, 40
Fleming, Peter, 17–18, 221–224, 234
Flexibility, employee, 241
Flyverbom, M., 166–167, 207–209
Forbes Magazine, 251
Ford, Henry, 32, 114
Fordists, assembly line and, 137
Formal organizational structure, 83
Foucault, Michel, 63, 224, 226, 234, 235
Fournier, Valérie, 109
Foxxconn, 125
Freelance workers
  empowerment and, 217–218
  power dynamics and, 220
  UpWork platform, 218–220
Freire, Paulo, 229
Friedman, Milton, 198
Fuchs, Christian, 124, 125

INDEX

G

Gantt charts, 92
General and Industrial Management (Fayol), 3
Gergen, Kenneth, 44
German Pirate Party, 212–213
Gibson, James, 68, 71
Glass ceiling, 222
Global North, 126
Global South, 126
Golden Age of Capitalism, 110–111
Google, 124, 231, 243, 251, 252, 253
Google Flu Trends, 209
Google Page Rank, 73
GPS data, 165
Gray, Barbara, 136
Grey, Chris, 109
Grint, Keith, 47, 65
Grolin, Jesper, 199
Groupware, 140
Guardian, The, 253
Gulbrandsen, I. T., 179, 181, 184–185
Gulick, Luther, 30
Gupta, M., 13

H

Hackathon, 149
Hacking, 248
Hansen, H. K., 166–167, 207–209
Hardt, M., 249
Hatch, Mary Jo, 40
Hawthorne Works studies, 34–35, 49, 109
Heilbroner, Robert, 56
Here Comes Everybody: The power of organizing without organization (Shirky), 230
HEROes (highly empowered and resourceful operatives), 231–232
Hierarchical organizations, 95, 102
Hierarchical relationships, 97–98
Historical materialism, 55–56
Holocracy, 92
Home care practice, knowledge, datafication and, 154–156
Hong, Sun-ha, 168
Hopper, Grace, 5
Howe, Jeff, 60, 230
H&R Block, 167
Huffington, Arianna, 24, 106–109
Huffington Post, 24, 106–108

Human knowledge, learning and, 158
Human organization, 34
Human Problems of an Industrial Civilization, The (Mayo), 49
Human Relations School, 33–35, 39
Husted, E., 249, 250
Hutchby, Ian, 69, 71
HyperText Markup Language (HTLM), 12
HyperText Transfer Protocol (HTTP), 12

I

IaaS (Infrastructure as a Service), 97
Iansiti, M., 13
IBM, 13, 158, 167
Identity, 224
Identity regulation, 224
Ideological bias, 197
Ideological domination, 224
Ikea, relevance and interactive communication, 188–191
Individualized production, 116–117
Individualized work, 116
Individual rewards, produsage and, 120
Individuals, digital organizing impact on, 254
 deskilling/reskilling, 240–242
 surveilled/measured, 243–244
 work-life balance, 242–243
Industrial age, to post-industrial age, prosumers and, 113–115
Industrial economy
 knowledge and, 115
 socialized work and, 116
Industrialization, 108
Industrial Organisation (Woodward), 37
Industrial revolution, 114
Industry 4.0, 16
Information, transparency needs more/better, 204–208
Information Age, The (Castells), 115
Informational economy
 individualized work, 116
 production and, 115–116
Information Management Science, 96
Information Society, 246
Information technology (IT)
 industrial civilization and, 59–60
 infrastructure, 96–98
 manufacturing and, 116–117
 networked organization and, 90

263

INDEX

Infrastructures, organizational
  IaaS (infrastructure as a service), 97
  information technology (IT) and, 96–98
  reorganizing, mental health, 99–100
  social media and, 98–99
  sociomateriality and, 100–102
  See also Organizational structure/infrastructure
Innosite (crowdsourcing platform), 146–148
Innovation
  collaboration and, 138
  as disruptive, 16
Innovator's Dilemma (Christensen), 16
Instagram, 186, 189
Institutional approaches, organizational legitimacy, 199–201
Intangible content production, 126
Intel, 9
Interactivity, 22
Interactivity, communication and
  communication theory and, 185–186
  complexity, 184–185
  digital technologies and, 244
  media theory and, 186–187
  organizations and, 187–188
  sociomateriality and, 191–192
  See also Communication, interactivity and
Internal communication, 178
International Consortium of Investigative Journalists, 131
Internet
  content produsage and, 126
  cyber proletariat and, 125–126
  development of, 11–12
  economic surveillance and, 124
  ethos of openness and, 12–14
  interactivity, communication and, 186–187
  languages of, 12
  passive consumer and, 117–118
  user-generated content and, 123
  web-based technologies, collaboration, 140–141
  Wikipedia, 3D printing and, 119–122
Internet of Things (IoT), 14
Investigation of the Laws of Thought, An (Boole), 4
iPhone, 16, 60, 110, 112
iPod, 144–145
IT Enterprise Systems, 97

J
Jeopardy (game show), 158
Jobs, bounded automation and, 17–19
Just, S., 179, 181, 184–185
Justesen, L., 154n1, 162n1, 247n1

K
Kamstrup, A., 146–148
Kanter, R. M., 222
Kaplan, Sarah, 69
Kelly, Kevin, 60–61
Keynes, John Maynard, 17
Knights Capital Group, 45
Knowledge
  big data and, 166–168
  datafication, home care practice, 154–156
  data relationship, 163
  generation of, social practices, 137
  industrial economy and, 115
  organizational learning, 157–158
  organizational studies and, 156–157
  real-time data into actionable, 168–169
  sociomateriality, datafication and, 170–171
Knowledge Application Systems, 161
Knowledge-assets, 160
Knowledge-based firms, networked organization, 91
Knowledge Capture Systems, 160–161
Knowledge Discovery Systems, 160
Knowledge-intensive environment, collaboration and, 138
Knowledge management
  defining, scholars on, 159–160
  systems (KMS), 160–162
Knowledge Sharing Systems, 161
Knowledge society, 153, 160
Knowledge work, 137
Knowledge worker, 157
Kornberger, Martin, 99
Koushik, S., 190
Kreiner, K., 135
Kunda, G., 224

L
Labor
  assembly line, 57, 137, 225–226
  cheap, exploitation of, 253
  division of, 137
  See also Manual workers

INDEX

Labor and Monopoly Capital: The Degradation of Work in the Twentieth Century (Braverman), 111
Laboratory collaboration, 138
Labor process theory, 110–113
Labor theory of value, 110
Labor value, 110–111
Laclau, Ernesto, 64
Lakhani, K. R., 13
Language, Internet technologies and, 12
Lanzolla, G., 246
Large-batch production firms, 37, 41–42
Latour, Bruno, 65–66
Lave, J., 137
Law, John, 66–67
Lawrence, Paul, 40, 42, 49
Lazar, David, 209
Leadbetter, Charles, 60
Learning platform, features of, 206
Legitimacy
  collective negotiation, 200
  corporate models of, 198
  Corporate Social Responsibility (CSR) and, 201
  evaluation process, institutional approach, 199–200
  gaining/maintaining, 201–202
  goal of, strategic approach, 201–203
  political corporation model of, 199
  repairing, 202
  sociomateriality, transparency and, 212–214
  Stakeholder Theory and, 203–204
Legitimacy, transparency and
  defining, need for, 196–197
  Petrobras oil company, case, 197–198
  pharmaceutical company example, 213
LEGO company, 93
Leonardi, Paul, 69–70, 161, 189
Lepore, Jill, 16
Long Island bridges, 61–63
Lorsch, Jay, 40, 42, 49
Los Angeles Police Department, 169
Lukes, Steven, 223
Lyotard, Jean-François, 43, 49

M

Machine bureaucracy, 87
Machine learning, artificial intelligence and, 10
Machines
  algorithmic thinking and, 4–5
  decision-making, delegating to, 10–11
  driverless cars, 10–11
Madsen, Anders Koed, 70–71
Management 2.0, 94, 190
Managers, workers *v.*, digital technology and, 226
Manipulation, power and, 221–222, 234
Manual workers
  Chinese factory, 126
  enforced standardization, 31–32
  exploitation of, 111, 112, 124–125
  production and, 108–109
  technology and, 112–113
  theory of scientific management and, 109
  See also Labor
Manufacturing
  computer-aided systems of, 114
  for exchange, 114
  organizational transparency, 208–210
  smart machines and, 225–226
Marx, Karl, 54–62, 110–112, 116, 125–126
Marx, Leo, 58
Marxism, organization technology and
  pessimism/optimism, 57–61
  productive forces, 55–56
  technology-dependent alienation and, 57
  theory of historical materialism, 55
Mass media, influence of, 186
Mass-production firms, 37
Material approach, organizational communication, 183–184
Materialism, historical, 55–56
Matrix, The (film), 54
May, Theresa, 251
Mayo, Elton, 34, 49
Mazmanian, M., 242–243
McAfee, Andrew, 10–11
McKinsey Global Institute, 17
Meaning, communication co-production and, 185
Meaning-making, communication and, 184
Measured individual, surveilled individual *v.*, 243–244
Mechanization, 38
Media studies, 186–187

265

INDEX

Media theory, communication interactivity and, 186–187
Mental health, infrastructure reorganization, 99–100
Metanarratives, 44
#MeToo movement, counter-power and, 233–235
Meyer, J. W., 223
Michels, Robert, 221
Microchips
Millo, Yuval, 19
Mintzberg, Henry, 87
Mobile devices
 interactivity and, 187
 work-life balance, 243
Models of corporate legitimacy, 199
Modern structuralism, 86–87
Modern Times (film), 32
Modi, Narendra, 251
Moneyball (film), 164
Moore, Gordon, 9–10
Moore, M., 253
Moore's law, 9–10
Morgan, Gareth, 43
Morozov, E., 169
Moses, Robert, 58–59
Mouffe, Chantal, 64
MoveOn, 250
Mumby, D., 181–182
MySpace, 117, 123, 250

N

Narrative communications, 183–184
Negri, A., 249
Netflix, 84
Network connectivity, 11–12
Networked organization, 89–92
 digital technologies and, 92
 examples of, 91
 shared office spaces, 91
Network enterprise, 116, 117
New Left activists, 59
New technology, 44–45
New York Metropolitan Transit Authority, 61–62
New York Police Department (NYPD), 169, 230
New York Stock Exchange (NYSE), 19, 20, 143
Nike, co-creation and, 144–146
NikePlus, 144–146

Nonaka, Ikujiro, 115–116, 159, 160
Non-decision-making, 222
Non-governmental organizations (NGOs), 245, 248–249
Non-routine technologies, 42
Novo Nordisk, 181
Nyberg, Daniel, 127

O

Occupy Wall Street, 100–102, 249
Oligarchization, power and, 249
Olivia (magazine), 132
Olson, M., 134
O'Neil Cathy, 8–9, 168
Online collaboration, 94–96
Online production, 118
Online users, user-generated content, 123
Open Access, 246
Open Innovation, 246
Openness
 co-creation and, 142–144
 ethos of, 12–14
Open participation, produsage and, 119–120
Open Strategy, 246
Open system, organization as, 41
Optimism, technological, 57–61
Organization(s)
 artifacts in, 184
 big data and, 165–166
 binary logic, digitization-ready legislation and, 5–7
 co-creation and, 132–133
 collaboration between, 131–132
 connectivity and, 11–15
 data-driven decision-making and, 162–164, 170
 digitalization of, 7–10, 245–246
 digital technologies, collaboration within, 141
 exponentiality, digitalization and, 9–10
 knowledge datafication in, 154–156
 knowledge management and, 160
 manufacturing transparency, 208–210
 openness paradigm, co-creation and, 142–144
 as open system, 41
 private firms, digitalization and, 246–247
 public, digitalization and, 247–248
 self-organizing teams, 115

266

INDEX

social media and, 70
technology and. *See* Technology, organization and
voluntary, digitalization and, 248–250
Organizational boundaries, 189
Organizational communication
  interactivity and, 187–188
  perspectives/core concepts, 191
Organizational democratization, 39
Organizational design, 84–85
Organizational forms, structure and, 87–88
Organizational identities, 250
Organizational knowledge, 157
Organizational learning
  defining, 157–158
  human knowledge/learning, Watson case, 158
Organizational performance, rankings/benchmarks, 208–209
Organizational structure/infrastructure
  classical organization theory and, 81–82
  digital infrastructures, 83
  digital organizing, virtual library, 82–83
  formal organizational structure, 83
  production systems and, 36–37
  structure, studies of, 83–89. *See also* Structure, organizational
  technology and, 37
Organizational studies
  collaboration and, 133–141. *See also* Collaboration
  communication as theme of, 176–178
  contingency, 39–43
  equivocality and, 44–46
  efficiency and, 30–32
  Human Relations School and, 33–35
  knowledge and, 156–157
  legitimacy as theme, 197–204. *See also* Legitimacy, organizational studies and
  postmodernist, 43–44, 46
  sociotechnical theory, 37–38
  structure and, 83–89. *See also* Structure, organizational
  technology and, 36–37, 47–49
  theorists/contributors, 49
Organizational technologies
  craft, 42
  engineering, 42
  non-routine, 42
  routine, 41–42

Organizational thinking
  blended automation and, 19–21
  bounded automation, 17–19
Organization and Environment (Lawrence & Lorsch), 42, 49
Organization technology. *See* Marxism, organization technology and
Organizing, communication and, 183
Organizing production, 109
Orlikowski, Wanda, 47, 210–212
Ostrom, E., 135
Oxford Martin School, 17

P

Page, Larry, 251
Panama Papers scandal, 205
Panopticon, 226–227
Panoptic power, 226
Passive consumer, 117, 122
Peer production, 120
Peer-to-peer replication, 13. *See also* Blockchain *entries*
Performance, rankings/benchmarks and, 208–209
Perrow, Charles, 36, 40, 41
Persistence affordance, 70
Personal knowledge, 157
Perspectives, communication core concepts, 191
Pessimism, technological, 57–61
Petrobras oil, legitimacy/transparency, 197–198
Physical context, collaboration, 137–139
Pinterest, 189
Plant, Sadie, 4
Platform manager, 148
Plesner, U., 154n1, 162n1, 235, 247n1, 250
Podio, 139
Polanyi, Michael, 157
Police, big data non-transparency in work, 169
Political action, digital organizing, 249–250
Political corporation model, corporate legitimacy, 199
Political movements, digitalization of, 249–250
Political parties, digital organizing, 249–250
Political science, collaboration and, 135

267

INDEX

Poole, M. S., 177
Porter, Michael, 201
Postindustrial societies, 156
Postindustrial work, 137
Postmodern Condition, The (Lyotard), 49
Postmodern view, organizational studies, 43–44, 46
Powell, W. W., 223
Power
   automation, smart machines and, 225–226
   coercion and, 220–221
   counter-power and, 232–233
   defining, 220
   digital empowerment, 229–232
   discipline/informate, tax administration case, 226–229
   domination and, 222–223
   four faces of in organizations, 234
   individuals, society and, 251
   manipulation and, 221–222
   oligarchization and, 249
   sociomateriality, empowerment and, 235
   subjectification and, 223–225
   tech companies and, 251–252
Power: A Radical View (Lukes), 223
PowerPoint, knowledge production and, 69
Predictions
   behavior/consumption, 243
   data analysis and, 165–166
Predictive policing, 169
PredPol system, 169
Preindustrial society, 114
PricewaterhouseCooper, 163
Prilla, M., 140
Principles of Scientific Management, The (Taylor), 30, 31, 49
Privacy International, 245
Private firms, digitalization and, 246–247
Private organizations
   co-creation and, 144–146
   structure and, 92–94
Process production firms, 37
Process view, organizations, 44
Production, organizational studies
   conception/execution of, 112
   environmental cost, digital technology, 125–126
   exploitative nature of, 111, 112
   feedback loop, consumers and, 118

   intangible content, 126
   labor process theory, 110–113
   manufacturing for exchange and, 114
   online content, 118
   prosumers and, 113–117. *See also* Prosumers, production and
   sociomateriality, produsage and, 126–127
   themes/theories of, 108–110
   virtual, 118
Production-consumption process, 118
Production systems
   organizational structure and, 36–37
Produsage, 22
   common property, individual rewards, 120
   critiques/blind spots of, 122–126
   cyber proletariat and, 125–126
   fluid heterarchy, ad hoc meritocracy, 120
   open participation, communal evaluation, 119–120
   practice of, Wikipedia, 3D printing and, 119–122
   process of, 119
   produsers, 122–123
   from prosumption to, 117–122
   sociomateriality, production and, 126–127
   unfinished artifacts, continuing process, 120
   user commodification, 123–125
Professional bureaucracy, 87
Project organizations, 91
Proletariat, 125–126
Prosumers, production and
   individualized production, 116–117
   industrial age/postindustrial, 113–115
   informational economy, production in, 115–116
   preindustrial society and, 114
   technology and, 114
Prosumption, to produsage, 117–122
   consumers and, 122
   process, 119
Proudhon, Pierre-Joseph, 58
Public organizations
   digitalization and, 247–248
   structure and, 92–94
Public Relations, 205
Public sector

INDEX

citizen in digitized, 244–245
co-creation and, 146
Punch cards, 6
Pyramide capitaliste, 111

Q

Qualitative due diligence, 208, 210
Quantitative rankings, 208, 210

R

Ramaswamy, V., 144–146
Rand, Ayn, 59
Rankings, organizational performance and, 208–209, 211
Raviola, E., 235
Receivers, 185
Reception studies, 185
Reorganizing infrastructures, 99–100
Reskilling, workers, 226, 240–242
Rice, A. K., 39
"RIP the Consumer" (Shirky), 117
Rise of the Network Society, The (Castells), 90
Roethlisberger, Fritz, 34
Routine technologies, 41–42
Rowan, B., 223
Royal Library, 149
Rule-of-thumb, production management and, 108

S

Schadler, T., 231, 232
Schmidt, K., 139–140
Science and Technology Studies (STS), 65–66
Scientific Management movement, 30–32
Scott, Susan, 47, 87, 210–212
Second Life (game), 118, 121, 125
Self-organizing teams, 115
Sensemaking, 44, 182–183
Sensor data, 165, 166
Shadow political structure, 222
Shapeways 3D printing, 121
Shared office spaces, networked organization, 91
Shirky, Clay, 60, 117, 230
Sikkink, Kathryn, 93
Silicon Valley, 138, 224
Simple structure, 87
Singularity, 54

Singularity University, 10
Situated learning, 137
Small-batch firms, 36–37
Smart machines, power and, 225–229
Smartphone production, 125
Smith, Adam, 110
Smith, Merrit Roe, 58
Smythe, Dallas, 123–124
Snowden, Edward, 205
Social constructivism, 59, 61–65
  affordance and, 72
  analysis, 73
  anti-essentialism and, 62–63
  collaboration and, 137
  discourse theory, 64
  physical world and, 64
  technicism, 65
  truths and, 63
Social context, collaboration, 137–139
Social media
  affordances, 192
  co-creation/collaboration and, 132
  customer engagement, 145–146
  environmental awareness and, 125
  interactivity and. *See* Communication, interactivity and
  IT infrastructure and, 98–99
  knowledge management systems (KMS) and, 161–162
  Occupy Wall Street and, 101–102
  organizational use of, 180
  in organizations, 70
  uprisings on, 233–234
Social organization, of work, 140
Social Psychology of Organizing (Weick), 44, 49
Social shaping of technology (SST), 49
Society, digital organizing, 250–254
Socioeconomic order, 56
Sociomateriality
  actor-network theory (ANT), 66–67
  affordance and, 72
  analysis, 73
  collaboration/co-creation and, 148–150
  communication, interactivity and, 191–192
  defining, 67
  knowledge, datafication and, 170–171
  legitimacy/transparency and, 212–214
  power/empowerment and, 235

269

INDEX

Sociomateriality (*Continued*)
   production, produsage and, 126–127
   Science and Technology Studies (STS) and, 65–66
   structure/infrastructure and, 100–102
Sociotechnical theory, 37–38
Software, collaborative, 139
Spaghetti code, 11
Specialization, 112
Spicer, A., 221–234
Stakeholder communications, 205
Stakeholder management, 203
Stakeholder model of corporate legitimacy, 198, 199
Stakeholder Theory, corporate legitimacy, 203–204
Standardization, 32
Starr, Susan Leigh, 83
Stochastic events, 45, 46
Stock trading, blended automation and, 19
Strategic approach, organizational legitimacy, 201–203
Strategic communication, 179–180
Strategy, datafication paradigm and, 171
Structure, organizational
   bureaucracy, 86–87
   contingency theory and, 84
   critiques of, 88–89
   design, 84–85
   infrastructures, 96–100. *See also* Infrastructures, organizational
   networked, 89–92
   organizational forms, 87–88
   private/public organizations and, 92–94
   sociomateriality, infrastructure and, 100–102
   virtual, 94–96
Structure. *See* Organizational structure/infrastructure
Structuring of Organizations, The (Mintzberg), 87
Subjectification
   coersion, HEROes case, 231–232
   power and, 223–225, 234, 246
Suchman, Mark, 200, 201
Surplus value, 110, 111
Surveillance capitalism, 252
Surveillance technology, 206–207, 245
Surveilled individual, measured individual *v.*, 243–244
Sweatshop workers, 125

Symbolic-interpretive school, 44
Systematic soldiering, 108

T

Tacit knowledge, 115, 157, 159, 160
Tapscott, Don, 60, 117–118, 122
Target groups, 185
Tasini, Jonathan, 106–108
Task analyzability, 42
Task-related technology, 41
Task variability, 42
Tavistock Institute of Human Relations, 37–40, 49
Tax administration, power/discipline case, 226–229
Taylor, Frederick Winslow, 30–32, 49, 108–109
Taylorism, 32, 109
Teams, self-organizing, 115
Tech companies, power and, 251–252
Technical design, functionality and, 62
Technical organization, 34
Technicism, 65
Technicum, 60
Technological determinism
   affordance and, 72
   analysis, 72–73
   big data, knowledge and, 167
   materialism as expression of, 54–55
Technological optimism, 57–61
Technological pessimism, 57–61
Technology
   actor-network (ANT)/social shaping (SST), 49
   big data, police work and, 169
   Computer-Supported Cooperative Work (CSCW), 139–140
   deconstruction of, 47
   defined, 41
   knowledge management systems (KMS) and, 160–161
   mediating, transparency and, 208–212. *See also* Transparency, mediating technology and
   new, 44–45
   organizational, 41. *See also* Organizational technologies, 47–49
   power, smart machines and, 225–229
   produsage and, 120
   prosumer and, 114
   socioeconomic order and, 56

INDEX

web-based, collaboration and, 140–141
workers and, 112–113
See also Digital technology
Technology, organization and
affordances, 68–73
Marxism and, 55–61. *See also* Marxism, organization technology and
perspectives, 53–54, 73–74
social constructivism, 61–65
sociomateriality, 65–68
technological determinism, 54–55
Techno-optimist literature, 230, 231
Terminology, classical/contemporary, 21–22
Time-and-motion studies, 31–32
Theory of alienation, 112
Theory of scientific management
capitalism and, 111
manual workers and, 109
movement of, 32
Third Wave, The (Toffler), 59–60, 113, 117
3D printing, Wikipedia and, 119–122
Time Magazine, 117
Toffler, Alvin, 59–60, 113–115, 117, 122
Traditional organizational design, 84–85
Transaction technology, 13. *See also* Blockchain *entries*
Transistor development, 9–10
Transparency, legitimacy and
digital learning platform, more/better information and, 206–208
German Pirate Party, 212–213
information, more/better, 204–205
sociomateriality and, 212–214
Transparency, mediating technology and
organizational manufacturing of, 208–210
unwanted exposure, TripAdvisor case, 210–212
Transparency International, 204
Treem, Jeffrey, 69–70, 161
TripAdvisor, transparency case, 210–212
Trist, Eric, 38–39
Trump, Donald, 188
Turco, Catherine, 88–89
Twelpforce, 231–232
Twelve Principles of Efficiency, The (Emerson), 31
Twitter, 72, 175–176, 187–188, 191–192, 203–204, 249
Twitter Revolutions, 233n4

U

Unfinished artifacts, produsage and, 120
Uniform Resource Locator (URL), 12
Unpredictability, 45
User behavior, 123
User commodification, produsage and, 123–125
User-generated content, 123

V

Vaara, E., 200
Value creation, 144
van Dijck, José, 123
Variable geometry, 91
Vertically disintegrated organizations, 95
Villadsen, K., 242
Virtualization, 88
Virtual organization, 94–96
Virtual world, 118
Visualizations
affordances and, 70
data displays, 166
Volkswagen, 202
Voluntary organizations, digitalization of, 248–250

W

Wajcman, J., 243
Watson, IBM AI project, 158, 167
Weapons of Math Destruction (O'Neil), 8
Web 2.0, 140, 161, 186
Web 3.0, 161
Web-based technologies
collaboration, 140–141
knowledge management, 161
Weber, Max, 33, 86, 220–221
Weick, Karl E., 44–46, 49, 182–183
Wenger, E., 137
Western Electric Company, 34
Western Marxism, 124
What Technology Wants (Kelly), 60
Whistleblowers, 204–205
WikiLeaks Revolution, 233n4
Wikinomics, 117, 122
Wikipedia, 117, 119–122, 125
Wikis, 139
Williams, Anthony, 60, 117–118, 122
Willmott, Hugh, 64, 224
Winner, Langdon, 58–59

INDEX

Wired magazine, 60, 120, 166
Woodward, Joan, 36–37, 39, 49, 56, 57
Woolgar, Steve, 47, 61–62, 65
Work, social organization of, 140. *See also* Labor; Manual labor
Workflow data, 166
Work-life balance, digital organizing and, 242–243
World Wide Web, 12

Y

YouTube, 117, 118, 186, 189, 218–219, 231, 250

Z

Zammuto, Raymond, 47, 70, 97
Zuboff, Shoshana, 97, 225–226, 228, 229, 235, 240–241, 243, 252
Zuckerberg, Mark, 251